Immunoassays in Coagulation Testing

Zaheer Parvez

Immunoassays in Coagulation Testing

With 50 Figures

Springer-Verlag
New York   Berlin   Heidelberg   Tokyo

Zaheer Parvez, Ph.D.
Department of Radiology and Pathology
Loyola University Medical Center
Stritch School of Medicine
Maywood, Illinois, U.S.A.

Library of Congress Cataloging in Publication data
Parvez, Zaheer.
  Immunoassays in coagulation testing.
  Bibliography: p.
  Includes index.
  1. Blood coagulation tests.   2. Blood coagulation
factors—Analysis.   3. Immunoassay.   I. Title.
RB45.P37   1984   616.07′561   83-20115

Typeset by Bi-Comp, Incorporated, York, Pennsylvania.

ISBN 978-1-4615-7227-5   ISBN 978-1-4615-7225-1 (eBook)
DOI 10.1007/978-1-4615-7225-1

9 8 7 6 5 4 3 2 1

In loving memory of my mother

This book is respectfully dedicated to those
who laid the foundations of immunology in coagulation,
and
to my wife, Kausar, and my children, Farah and Fareen.

Foreword

Rapid progress in analytical methods, within the past few decades, has led to the widespread applications of newer immunological and radioimmunoassay techniques to the diagnosis and treatment of hemorrhagic and thrombotic disorders. Major advances were made to meet the multiple challenges of improving precision, accuracy, and availability of various measurements. These advances have been paralleled by the discoveries of a close relationship between biological activities and the absolute concentration of proteins that were measured by immunological techniques. This, in turn, assured the significance and usefulness of immunological methods in the management of patients. Numerous variants of immunological tests now are available, which allow us to both determine with precision minute quantities of antigenic proteins in serum and other biological fluids and differentiate the native protein from its genetically altered or degraded forms. Methods also have been designed to immunologically evaluate some serine proteases that are in complex with proteolytic inhibitors. Due to rapid progress in this field, different laboratories unavoidably become experts in one or the other approach. In the welter of possible choices, the non-expert usually is left to follow either the most recent but as yet to be confirmed method or his own anecdotal experience. This manual not only brought together various methods in current use, but it also set certain standardized criteria for the assessment of various deficiencies and abnormalities in hemostasis.

Chapters 1 and 2 deal with basic concepts in coagulation and the pathophysiology of coagulation disorders, which merely serve the purpose of acquainting the reader with their interrelationship; persons with advanced knowledge may wish to skip this discussion. Chapter 3 deals with the basic terminology, concepts of antigens, antigenic determinant sites, antibodies, their binding, and specificity, and poly and monoclonal antibodies. Some aspects of immunization techniques that are fundamental problems in immunology are reported in the Appendix. Chapter 4 reviews the immunological techniques presently in use. The principles of the methods are explained by photographs and diagrams with the purpose of using this book as a laboratory manual. Some readers with a knowledge of certain routine laboratory techniques will be acquainted with the use of more advanced methodologies, which are less frequently used. Helpful comments on their potential use and clinical interpretation also are provided. Chapter 5 includes immunoassay procedures used for some proteins of the coagulation and fibrinolysis systems. Two other components of hemostasis, mainly serine protease inhibitors and two platelet specific proteins (PF4 and BTG), due to their major importance in a clinical diagnosis are also covered in this chapter. Nevertheless, due to the logarithmic growth of this discipline, some areas remained open-ended. The neglected area, which could be covered by additional appendix material in successive editions, includes certain proteins of the complement systems as well as the influence of interfering lipids and glycolipids, which are assumed to have a particular importance in the interaction of coagulation factors with their antibodies. Chapter 6 touches on the research applications of immunological techniques in isolating and purifying clotting factors and inhibitors. Finally, in Chapter 7, the problems associated with quality control are described. In the Appendix, not only the normal ranges of coagulation proteins in plasma are covered, but also a list of manufacturers of immunological reagents and equipments is provided.

Dr. Zaheer Parvez, Ph.D., the author who has challenged this task, has wide experience in immunology. His knowledge is derived from continuous applications of immunological techniques to hemorrhagic and thrombotic disease at Loyola University Medical Center of Chicago, under the auspices of Prof. H.L. Messmore. In assembling this work, Dr. Parvez has attempted to selectively bring together but with representative completeness, the series of thoughts, experimental studies, and clinical applications that have characterized the surging influence of immunological testing in hemostasis. It is hoped that many newcomers in the field will be acquainted with various possibilities and limitations as well as optimal uses of immunological methods.

This manual will provide many persons interested in the field of hemorrhagic and thrombotic diseases with an up-to-date treatment of the subject and will serve equally the needs of medical technologists, clinical trainees, and clinicians, whose daily work is an assessment of human disease with an improved methodology.

Dr. M.J. Seghatchian, B.SC., Ph.D.
Blood Product Quality Control
and Research and Development, N.L.B.T.C.
Edgware, Middlesex, United Kingdom

Honorary Senior Lecturer
Guy's Hospital and Medical School
London, United Kingdom
February, 1984

Preface

Since my association with the Loyola University Medical Center, I had the opportunity to participate in several national and international symposia and workshops on *Newer Methods in Coagulation Testing*. These workshops were held at the annual meetings of the American Association of Clinical Chemistry, American Society of Medical Technologists, American Society of Clinical Pathologists, and the Association of Clinical Scientists. A large number of participants had indicated a need for a "cookbook" type manual of immunologic procedures that are used in testing coagulation disorders. Considering this genuine interest of a cross-section of medical technologists, clinical chemists, pathologists, and allied health professionals, a decision was made to publish my workshop presentations on immunologic methods in coagulation testing in the form of a laboratory handbook.

This manual originally was designed to provide basic operational procedures of widely used immunotechniques, so that it could be used at the bench level by technologists. However, it soon was realized that a manual of this type will not properly serve those who need conceptual information in coagulation and immunology as well. Therefore, the first part of this manual deals with the biochemical mechanisms in blood coagulation, the pathophysiology of major coagulation disorders, concepts in immunology, and a review of methods. Nevertheless, an in-depth discussion of hemostasis, pathogenesis of thrombosis, platelets, and vascular function defects was beyond the scope of this manual. A great deal of emphasis, therefore, is given to the actual assay procedures, which are the central theme of this manual. Most of these procedures, especially rocket immunoelectrophoretic, crossed immunoelectrophoretic, laser nephelometric, and fluorometric immunoassays were developed and standardized in my laboratory. They are expected to provide accurate results when performed accordingly. Also, some of the radioimmunoassay procedures were derived from the information supplied by the manufacturers of assay kits; these sources are duly acknowledged. As this work is meant to be a practical guide rather than a comprehensive treatise on coagulation or immunology, an extensive bibliography has been purposely avoided. It is expected that this book will be received in the spirit in which it is written.

Zaheer Parvez
February, 1984

Acknowledgments

In the long history of mankind, some human beings were unkind, unappreciative, and revengeful; a few were tolerant, suffered persecution, and even prayed for the salvation of their enemies. At the time of writing this acknowledgment, I do not know to what category a future historian will classify me. However, I do know that I am grateful to Professor Rogelio Moncada (Director of Radiologic Education) and Professor Leon Love, (Chairman, Department of Radiology) for their constant encouragement, support—both moral and financial—and for appreciating this undertaking. I am equally indebted to my colleague and dear friend, Dr. Jawed Fareed (Director of Hemostasis Research Laboratories), and to Dr. Harry L. Messmore, Jr. (Professor of Medicine), for introducing me to coagulation and for their valuable suggestions, criticisms, and many discussions on this exciting area of coagulation immunology. I also am appreciative of the research opportunities and the philosophy of academic freedom promoted by Dr. John R. Tobin (Dean, Stritch School of Medicine) and Dr. Alexander G. Karczmar (Associate Dean). I am obliged to Dr. Edward Bermes, Jr. (Acting Chairman, Department of Pathology) for his interest and encouragement during the completion of this project.

The unqualified assistance of Maureen Havey and Robert Vick in making diagrammatic illustrations and taking photographs as well as the skillful typing of Kathy Baronak and Marilyn Shanahan is much appreciated. Finally, I would like to extend my sincere thanks to the editorial staff of Springer-Verlag for their cooperation in the preparation of this manuscript.

Contents

Introduction

Since the earliest developments of quantitative precipitation and single radial immunodiffusion techniques,[44,73] clinical immunology has played a significant role in the study of infectious disease,[33,88,100] blood group antigens,[1,98,117,122] serum immunoglobulins,[3,9,16,51,99] detection of hormones,[4,21,27,30,49] drugs,[17,65,83] and histocompatibility mechanisms.[12,15,111] However, quantitative immunoassays were not effectively utilized in testing coagulation disorders until recently. The probable cause for such delay has been the late discoveries of coagulation factors and subsequent elucidation of biochemical mechanisms of coagulation.[23,55,69,77,93,96] In recent years, the precise role of various serine proteases and their inhibitors in the regulation of hemostatic functions has been determined[43,45,75,80] and amino acid sequences of thrombin, prothrombin, fibrinogen, and antithrombin-III discovered.[60,70,87] The physiology of hemostasis is now explained in relation to the coagulation factors, activators, and inhibitors that are present in plasma. Among the serine protease inhibitors, antithrombin-III (AT-III) modulates the biologic functions of thrombin, factor Xa, plasmin, and kallikrein, thereby regulating the formation of a blood clot. A genetic deficiency of AT-III has been shown to cause thrombotic episodes.[10,61,97,101,106] In addition, molecular abnormalities of AT-III in patients have also been recognized, and in such patients, functional activity of AT-III did not correlate to its immunologic concentration.

Similarly, α_2-antiplasmin, which is a primary inhibitor of the fibrinolytic enzyme plasmin, has been demonstrated to have more than one molecular form.[18,56] Molecular variants of coagulation factors, such as fibrinogen (87 variants), prothrombin (90 variants), Factor X, and Factor VIII, have also been reported.[114] Such clinical situations pose a serious challenge to our current diagnostic techniques, and hence, both functional and immunologic aberrations in coagulation proteins are to be determined. In a coagulation laboratory, clotting factors and serine protease inhibitors are quantitated by various functional and immunologic methods (Table 1). Although sound biochemical guidelines are available for functional assay procedures,[112,115] there is no single manual which

Table 1. Quantitative Methods in Coagulation Testing

Protein	Method
Fibrinogen	Clotting assay, EID, radial immunodiffusion
Prothrombin	Chromogenic substrate assay, EID
Factor VIII : RAg	EID, FIA, IRMA
Factor VIII : CAg	Immunoradiometric assay (IRMA)
Factor VIII c	RID, RIA, IRMA
Factor X	Chromogenic substrate assay, EID
Factor XII	Clotting assay, EID
Prekallikrein	Chromogenic substrate assay, EID
Plasminogen	Caseinolytic, chromogenic/fluorogenic substrate assay, RID, EID
Antithrombin-III (AT-III)	Clotting assay, chromogenic/fluorogenic substrate assay, RID, EID, RIA, EIA, NIA
α_1-Antitrypsin (α_1-AT)	Chromogenic substrate assay, RID, EID, NIA
α_2-Macroglobulin (α_2-M)	RID, nephelometry
α_2-Antiplasmin (α_2-AP)	Chromogenic substrate assay, caseinolytic assay, EID
Fibrin/fibrinogen degradation products (FDP)	Latex agglutination
Fibrinopeptide A (FPA)	RIA, EIA
Bβ 15-42	RIA
Platelet factor 4 (PF4)	RIA
B-Thromboglobulin (BTG)	RIA

EIA = enzyme immunoassay; EID = electroimmunodiffusion; RIA = radioimmunoassay; FIA = fluoroimmunoassay

provides essential information on newly emerging quantitative immuno-chemical techniques in coagulation testing which can be utilized by both clinicians and research workers. However, a recent publication[116] on the application of immunoassays in endocrinology, pharmacology, toxicology, allergy, autoimmune disease, bacteriology, tissue typing, mycology, plant virology, and veterinary medicine has devoted only eight pages to immunoassays in hematology.

The intent of this work is to provide basic theoretical background on isotopic and nonisotopic immunoassays, outline operational procedures, and discuss recent innovations in immunodiagnostic methods in coagulation testing. Since antisera to some vital coagulation proteins are not commercially available yet, a chapter is allocated to the techniques in antisera production. It is anticipated that such an assembly of technical information will bridge the gap that exists at this time, and will be found useful in immunodiagnosis of coagulation disorders.

1

Basic Concepts in Coagulation

The biochemical mechanisms in coagulation, kinin generation, and the fibrinolytic pathways are based on well-defined fundamental concepts, and as such, must be fully understood before attempting to examine the immunologic role of a given coagulation protein. In this section, terms such as hemostasis, vasoconstriction, platelet activation, coagulation cascade, and so on are discussed so that the interdependence of the components of each system becomes apparent and an appropriate testing procedure is utilized. Extensive reviews on biochemical mechanisms in coagulation and fibrinolysis have appeared in recent years and readers are referred to these for further information.[6,15,22,82,102]

Hemostasis

Hemostasis refers to the maintenance of blood in its fluid form within the blood vessels, and involves blood platelets, vascular endothelium, clotting factors, clot lysing (fibrinolytic) enzymes, and inhibitors of the coagulation and fibrinolytic pathways (Fig. 1).

Vasoconstriction

When a blood vessel is injured, platelets (Fig. 2) adhere to the vessel wall, undergo a morphologic change, and initiate platelet release reaction. During the release reaction, platelets secrete ADP, serotonin, calcium, fibrinogen, platelet factor 4 (PF4), β-thromboglobulin (β-TG), prostaglandins G, H, E, D, and F, thromboxane A and B, and other biogenic amines. Immediate constriction of vascular smooth muscle takes place, resulting in a smaller wound site to be plugged by the platelets and red cells. At this point, the coagulation system has been activated by platelet

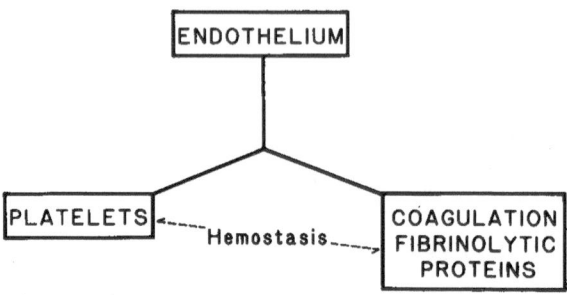

Figure 1. Components of hemostatic pathways.

phospholipids and other tissue thromboplastins, leading to the formation of fibrin from fibrinogen. Fibrin deposition at the site of injury reinforces the primary platelet plug and a stable clot is thus formed.

Prostaglandins and Platelet Aggregation

Prostaglandins are unsaturated fatty acids with 20 carbon atoms in a chain and are found in various body tissues. They are identified by the letters A, B, D, E, and F. Nonesterified arachidonic acid (AA) is the

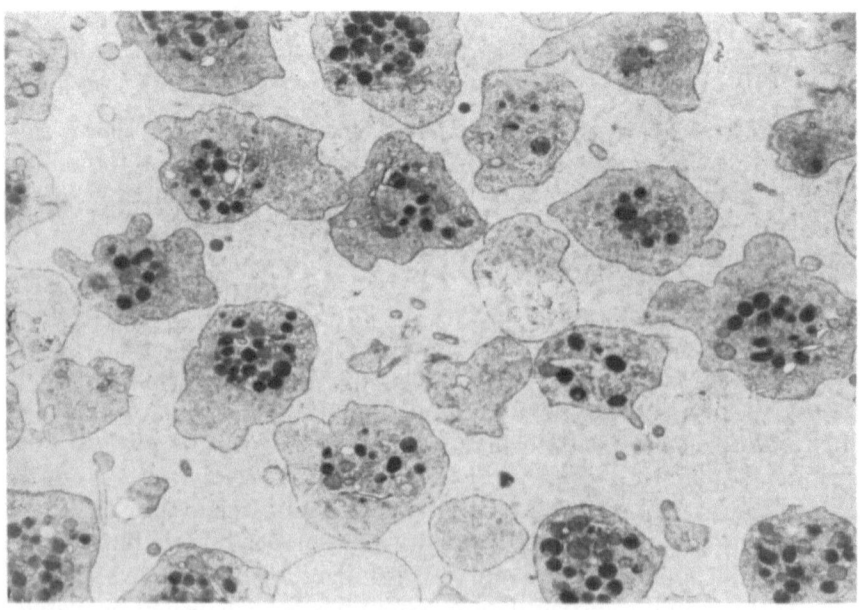

Figure 2. Electron micrograph of normal human platelets showing their granular contents. ×5000

precursor for prostaglandins and the first step in the production of prostaglandins is the release of the precursor from its esterified to its nonesterified (free) state. This function is achieved by the enzyme phospholipase A_2. The other lipolytic enzyme is a complex of two enzymes—cyclooxygenase and endoperoxide isomerase—commonly referred to as prostaglandin synthetase. The extensive work of Sune K. Bergstrom, Bengt I. Samuelsson, and John R. Vane was awarded with a Nobel Prize in 1982 for prostaglandin research in medicine. These workers described the biology and metabolic pathways of arachidonic acid transformation by human platelets.

Platelet aggregation is associated with metabolism of arachidonic acid to prostaglandin H_2 (PGH_2) through the cyclooxygenase pathway. Subsequently PGH_2 is converted to two main products, hydroxyheptadecatrienoic acid (HHT) and thromboxane A_2 (TXA_2). Thromboxane A_2 and its stable metabolite, thromboxane B_2 (TXB_2), induce platelet aggregation and arterial constriction. A different metabolite of PGH_2, called prostacyclin (PGI_2), is produced in the vessel wall and is known to inhibit platelet aggregation and relax smooth muscle. These two prostaglandins (TXA_2 and PGI_2) play an important physiologic role in regulating platelet aggregation and must be considered in evaluating hemostatic defects (Fig. 3).

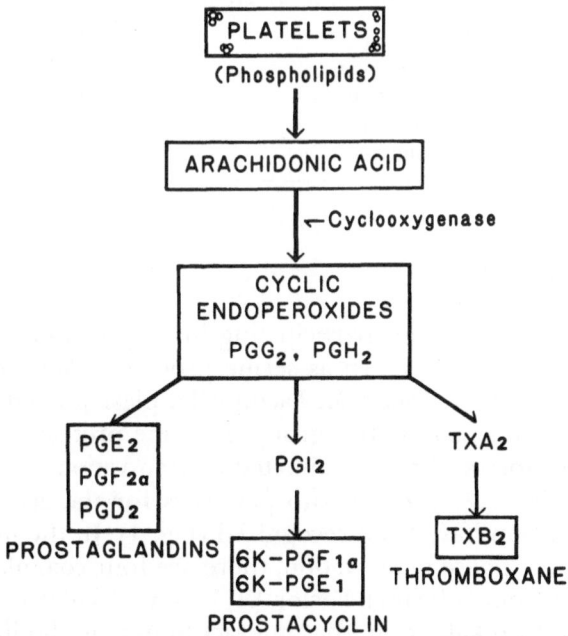

Figure 3. Regulation of platelet aggregation by prostaglandins and their metabolites. Thromboxane (TXA_2) stimulates platelet aggregation, whereas prostacyclin (PGI_2) inhibits such aggregation.

Table 2. Coagulation Factors in Plasma

Coagulation Factors	Synonyms	Molecular Weight	Concentration (μM)
Factor I	Fibrinogen	340,000	8.8
Factor II	Prothrombin	72,000	2.8
Factor III	Tissue factor	220,000	–
Factor IV	Calcium ion	–	–
Factor V	Proaccelerin	290,000	–
Factor VI	Activated V	–	–
Factor VII	Proconvertin	63,000	0.03
Factor VIII	Antihemophilic factor (AHF)	2,000,000	–
Factor IX	Christmas factor	55,400	0.60
Factor X	Stuart factor	55,000	0.13
Factor XI	Plasma thromboplastin antecedent (PTA)	160,000	0.043
Factor XII	Hageman factor	90,000	0.44
Factor XIII	Fibrin stabilizing factor (FSF)	320,000	–
Fletcher factor	Prekallikrein (PK)	88,000	0.34
Fitzgerald factor	High molecular weight kininogen (HMWK)	160,000	0.50
	Protein C	56,000	–
	Protein S	69,000	–
Plasminogen	Profibrinolysin	93,000	1.0
Fibronection	Cold insoluble globulin	450,000	–

Coagulation Cascade

Since most of the enzymes participating in coagulation act as trypsin, these enzymes are also known as serine proteases. Activated forms of Factor XII, Factor XI, Factor X, Factor VII, plasma kallikrein, thrombin, and plasmin constitute this group of serine proteases. A list of the coagulation factors and their estimated concentration in plasma is depicted in Table 2. At present, two pathways for the activation of the coagulation system have been recognized (Fig. 4). In the intrinsic pathway, i.e., contact phase of activation, there are four coagulation factors, Factor XII (Hageman factor), Factor XI, high molecular weight (HMW) kininogen, and prekallikrein, that interact to initiate the intrinsic pathway of coagulation, the fibrinolytic pathway, and the kinin-generating systems in plasma. Factor XII and kallikrein convert Factor XI to its

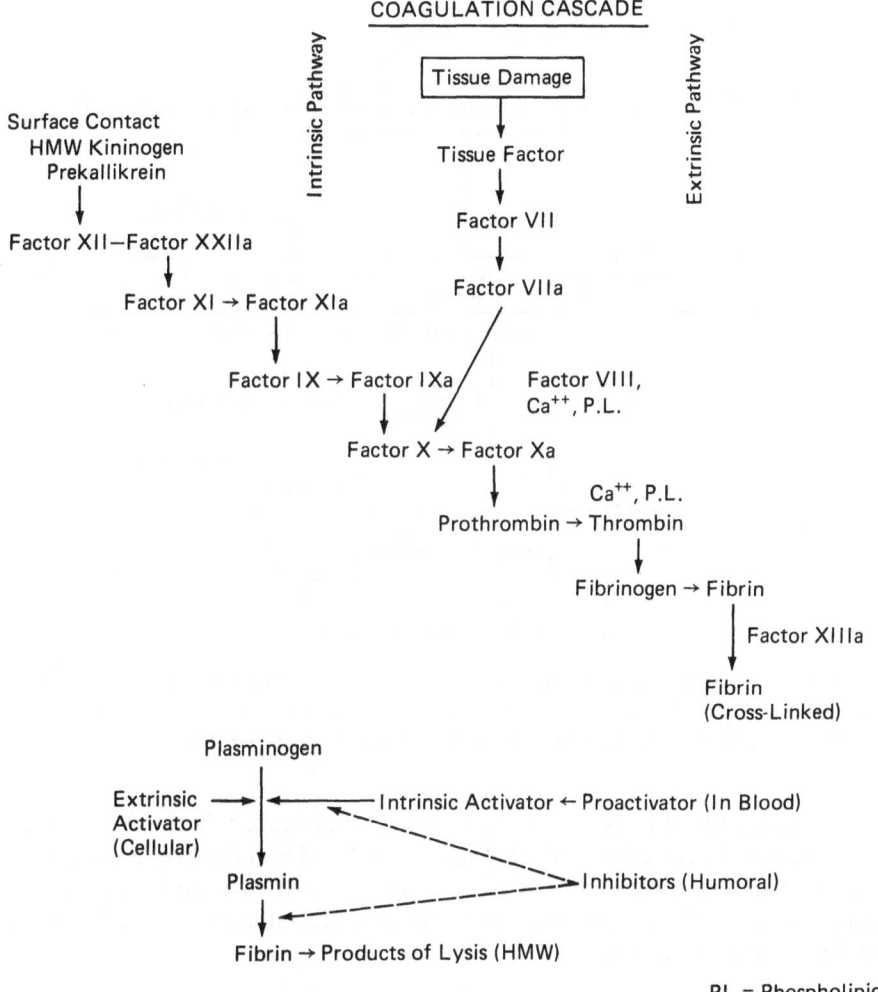

COAGULATION CASCADE

Figure 4. Molecular processes in a coagulation cascade.

active form, XIa. Factor XIa is capable of converting more Factor XII to XIIa, and leads to the activation of Factor IX. Factor X is then activated (Xa) by Factor IXa in the presence of Factor VIII, calcium, and phospholipids. A complex of Factor Xa–Factor V–calcium and phospholipid ultimately generates thrombin from its precursor, prothrombin. Once thrombin is generated, it transforms the soluble fibrinogen into an insoluble fibrin clot (Fig. 5).

The extrinsic pathway of coagulation is initiated via the activation of Factor VII by a lipoprotein called tissue factor, which is found in endo-

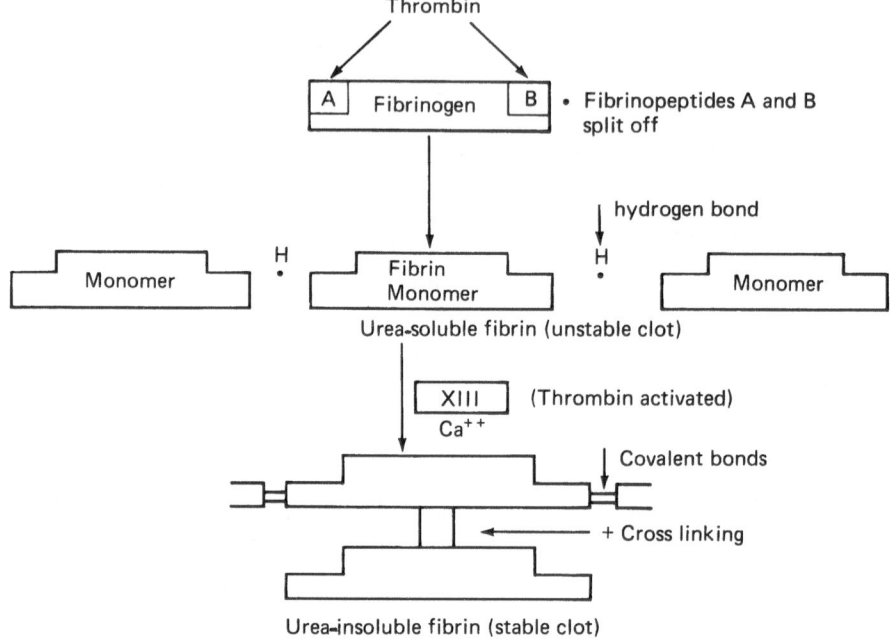

Figure 5. Thrombin cleaves fibrinogen into two molecules of fibrinopeptide A, two molecules of fibrinopeptide B, and fibrin monomers. These fibrin monomers are then cross-linked by Factor XIII to form a stable clot.

thelial cells, fibroblasts, and leukocytes. The activated Factor VII, along with Factor V, calcium, and platelet factor 3 (PF3), converts Factor X to Xa, which in turn acts on prothrombin to form thrombin. Thrombin interacts with fibrinogen to form fibrin and activates Factor XIII, which stabilizes the fibrin clot.

Fibrinolysis

Fibrinolysis is a physiologic process of dissolution of the formed blood clot. It is activated by the coagulation system or by the extrinsic activators. A plasma protein, plasminogen, and plasminogen activators and inhibitors are essential components of the fibrinolytic system. Plasminogen is activated to its enzymic form, plasmin, by activators of vascular and tissue origin. Urokinase and streptokinase are the two important activators of plasminogen. The active enzyme, plasmin, then breaks down fibrin(ogen) to its degradation products X, Y, D, and E, which act as inhibitors of thrombin and prevent further blood clotting (Fig. 6). Plasmin is also known to inactivate Factors V, VIII, IX, XI, and XII. The

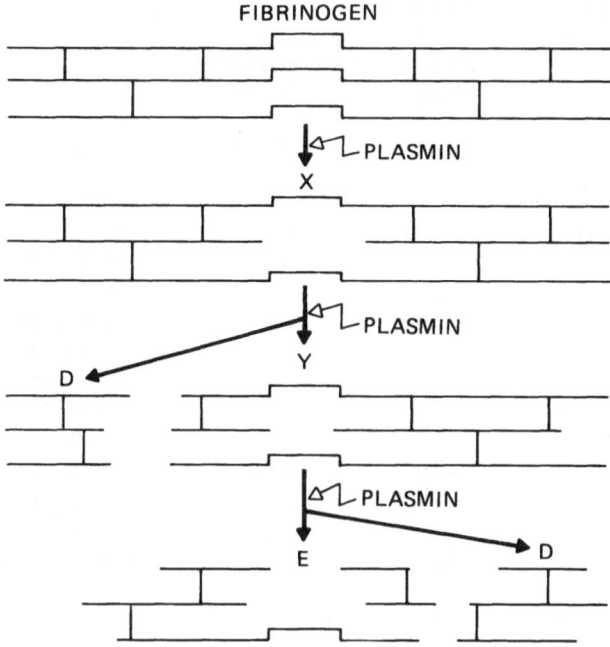

Figure 6. Proteolytic digestion of fibrinogen by plasmin, resulting in fibrinogen degradation products X, Y, D, and E.

excessive proteolyic activity of plasmin is controlled by a fast acting inhibitor, α_2-antiplasmin, and a slow reacting inhibitor, α_2-macroglobulin (α_2-M). A congenital deficiency of α_2-AP may result in a hemophilialike clinical bleeding disorder.

Serine Protease Inhibitors

The activation of coagulation and fibrinolytic systems has been shown to occur simultaneously so that the formed clot is lysed by the fibrinolytic enzyme, plasmin. The activities of coagulation and fibrinolytic enzymes are coordinated by the combined activity of serine protease inhibitors (Table 3), and more importantly, by antithrombin-III (AT-III) and α_2-antiplasmin (α_2-AP). The functional concentration of these inhibitors plays a critical role in anticoagulant, procoagulant, and thrombolytic therapy. Antithrombin-III is a progressive inhibitor of thrombin; however, in the presence of heparin, this inhibitor is activated and becomes a more potent inhibitor of Factor Xa and thrombin. The therapeutic usefulness of heparin mainly depends on the plasma concentration levels of AT-III, and patients with reduced levels of AT-III fail to respond to

Table 3. Physiologic Inhibitors of Coagulation Proteins

Inhibitor	Molecular Weight	Concentration (μM)	Enzymes Inhibited
α_1-Antitrypsin	55,000	35–71	Trypsin, chymotrypsin, plasmin, thrombin, collagenase, elastase
Antithrombin-III	65,000	3.5–6.3	Thrombin, kallikrein, Factor IXa, Factor Xa, Factor XIa, Factor XIIa
α_2-Macroglobulin	725,000	2.0–5.8	Trypsin, chymotrypsin, plasma, kallikrein, thrombin, plasmin, elastase
α_2-Antiplasmin	65,000	1.5–1.8	Plasmin
α_2-Antichymotrypsin	69,000	4.4–8.7	Chymotrypsin
Inter α-trypsin inhibitor	160,000	1.2–4.4	Trypsin, chymotrypsin (weak)
Cl inactivator	104,000	1.4–3.3	Factor XIa, Factor XIIa, kallikrein

heparin treatment. Some of the pathologic disorders in which AT-III concentration will be decreased are liver diseases, disseminated intravascular coagulation, deep venous thrombosis, acute leukemia, carcinoma, gram-negative septicemia, and fibrinolytic disorders. Laboratory monitoring of these inhibitors provides useful information in the management of hemostatic disorders. Since these inhibitors are mostly glycoproteins they are effective immunogens, and therefore, high titer monospecific antisera are readily available and may be utilized in both isotopic and nonisotopic immunoassays.

Suggested Readings

Bloom AL, Thomas DP (eds): Hemostasis and Thrombosis. New York, Churchill Livingstone, 1981.

Colman RW, Hirsh J, Marder VJ, Salzman EW (eds): Hemostasis and Thrombosis. Basic Principles and Clinical Practice. Philadelphia, Lippincott, 1982

2

Pathophysiology of Coagulation Disorders

Blood is maintained in its fluid state by complex interactions between endothelium, platelets, plasma proteins, and activators and inhibitors of the coagulation and fibrinolytic systems. This hemostatic balance can be altered by many undefined pathologic and physiologic processes resulting in coagulation disorders. In addition to these processes, diseases related to liver, kidney, heart, and lung functions, bacterial and viral infections, and malignancies are also known to cause coagulation abnormalities. Both inherited and acquired coagulation factor deficiencies result in coagulation defects, and in most cases, patients may not show clinical bleeding. As it is not possible to review all these coagulopathies in this treatise, a brief account of the pathophysiology of some major coagulation disorders will be presented in order to emphasize the critical role of immunoassays in detecting certain diagnostic parameters related to these disorders. For a detailed review on the pathophysiology of congenital coagulation defects the reader is referred to a recent publication.[71]

Hemophilia A (Classic Hemophilia)

Hemophilia A is a functional deficiency of antihemophilic factor or Factor VIII[91] and is characterized by a prolonged activated partial thromboplastin time and a normal prothrombin time. Partial thromboplastin time is prolonged since Factor VIII lacks the necessary coagulant activity. Intraarticular and deep muscle hemorrhages are clinical symptoms of the disease, which is inherited as a sex-linked recessive disorder and is manifested in males. Extensive studies by Shanberge and Gore[103] and Zimmerman and his colleagues[123,124] have shown that patients with classic hemophilia synthesize a nonfunctional form of antihemophilic factor devoid of clotting activity but capable of neutralizing antibodies against

Table 4. Sex-linked Recessive Inheritance Patterns
of Hemophilia

		Children	
Father	*Mother*	*Male*	*Female*
Hemophilic	Normal	100% Normal	100% Carrier
Normal	Carrier	50% Normal	50% Carrier
Hemophilic	Carrier	50% Hemophilic	50% Hemophilic

this protein. This antigenic protein is known as Factor VIII–related antigen, and is assayed for detecting carriers of hemophilia A. Therapy of classic hemophilia includes repeated transfusions of plasma, plasma fractions rich in antihemophilic factor (cryoprecipitates), or Factor VIII concentrates. The inheritance pattern of hemophilia is shown in Table 4.

Hemophilia B (Christmas Disease)

Hemophilia B was first discovered by Aggeler and associates[2] and subsequently by Biggs and her coworkers.[8] The disease is characterized by an abnormal activated partial thromboplastin time (Factor IX deficiency) and a normal prothrombin time. It is inherited as a sex-linked recessive character and is expressed in males. At present, hemophilia B patients are managed by infusing prothrombin complex concentrates, rich in Factor IX.

von Willebrand's Disease

Unlike classic hemophilia, von Willebrand's disease (vWD) is a hereditary disorder in which both functional and immunologic Factor VIII deficiency are present. The disease is characterized by easy bruising and mucosal bleeding. It is inherited as an autosomal dominant trait. The laboratory tests show prolonged bleeding time, abnormal platelet retention in a glass bead column, and impaired platelet aggregation with ristocetin.[46] The pathogenesis of vWD can be properly understood in the light of recent structural studies on antihemophilic factor (Factor VIII). It is now generally recognized that antihemophilic factor is a highly complex protein with a molecular weight of 1–2 million daltons,[50,64] and is composed of high molecular weight (HMW) and low molecular weight (LMW) protein subunits. The HMW subunit does not

have a procoagulant activity and is required for normal platelet retention in glass bead column and ristocetin-induced platelet aggregation. Its
concentration is greatly decreased in vWD, whereas in hemophilia A it
remains normal and constitutes a major portion of Factor VIII–related
antigen. The LMW subunit, on the contrary, has a procoagulant activity
and is lacking in hemophilia A and very low in vWD.

Disseminated Intravascular Coagulation (DIC)

DIC is the simultaneous activation of coagulation and fibrinolysis due to
several triggering mechanisms, such as infection, shock, amniotic fluid,
dead fetus, whole blood transfusion, malignancy, and several other stim-
uli.[7] A systemic generation of thrombin then leads to the formation of
microthrombi in the microcirculation. If the thrombin generation con-
tinues, the patient is likely to develop thrombosis. Alternatively, a subsequent activation of fibrinolysis gives rise to systemic plasmin, which at-
tacks fibrin(ogen) to yield fibrin(ogen) degradation products (FDP)—X,
Y, D, and E fragments. These fragments act as anticoagulants and, together with systemic plasmin, may cause serious hemorrhagic episodes.
DIC is diagnosed by elevated levels of circulating FDP, prolonged
thrombin time, partial thromboplastin time, and low levels of fibrinogen,
Factors V, VIII, II, and antithrombin-III. Clotting times in DIC are
prolonged because of hypofibrinogenemia, digestion of Factors V, VIII,
IX, and XI by plasmin, and the anticoagulant actions of FDP.

Platelet Function Defects

As pointed out earlier, platelets can adhere to collagen and other surfaces, and then aggregate in response to adenosine diphosphate (ADP),
epinephrine, or thrombin. The next step after adhesion is to synthesize
prostaglandin endoperoxides (PGG_2, PGH_2) and thromboxane A_2, and
finally undergo a release reaction. In some patients, the secretory process is greatly impaired and platelet aggregation does not occur. Based
upon this characteristic, platelet function defects are now classified as
inherited and acquired disorders. Inherited dysfunctions are due to
platelet membrane abnormalities, intracellular abnormalities, and mis-
cellaneous aberrations, although the nature of this inheritance is not
fully established. Acquired defects of platelet function are recognized in
myeloproliferative disorders, liver disease, alcoholism, drug therapy,
ischemic vascular disease, and diabetes mellitus. However, it is difficult
to ascertain the exact role of platelets in causing hemorrhagic or thrombotic complications associated with these pathologic conditions. Since

platelet immunology has been extensively investigated it will suffice to note that immunologic techniques have greatly contributed to the understanding of mechanisms in drug-induced thrombocytopenia, the nature of idiosyncrasy in patients, and the rationale for effective therapy.

Hypercoagulability

A hypercoagulable or prethrombotic state may be recognized as a tendency toward vascular clot formations, and as such constitutes an important parameter in coagulation testing. A hypercoagulable state in blood may arise due to an increased activity of platelets, resulting in arterial thrombosis, or due to elevated levels of certain clotting factors which may produce venous thrombosis. Increased levels of fibrinogen, prothrombin, factors V, VII, VIII, IX, X, and XI, as well as decreased concentration of AT-III, are usually associated with a prethrombotic state. Clinical situations which show a hypercoagulable state include carcinoma, myocardial infarction, congestive heart failure, cirrhosis, pregnancy, and the use of oral contraceptives.

Laboratory testing of hypercoagulability consists of determining AT-III levels, fibrinopeptide-A, FDP, and activated factor X. Radioimmunoassays can be utilized to detect fibrinopeptide A, FDP, and other fibrin monomer complexes.

Laboratory Testing

Although it is not feasible to describe here all the coagulation profiling tests used in a coagulation laboratory, a few important screening procedures will be discussed. Basic screening tests for primary hemostasis defects include Ivy bleeding time, platelet count, platelet adhesiveness, platelet aggregation, and clot retraction. Bleeding time measures the time for the blood to stop flowing from a 5 mm × 1 mm incision made on the forearm. In healthy individuals it ranges between 3 and 8 minutes and begins to prolong as platelet count drops below 100,000/mm^3 (platelet count in healthy individuals is 200,000–400,000/mm^3). In von Willebrand's disease bleeding time is prolonged, whereas in hemophilia A, the bleeding time is normal. Platelet adhesiveness, which is measured by running whole blood through a column of glass beads, is poor in vWD and normal in hemophilia A. Under normal circumstances, more than 25% of platelets adhere to glass beads. Platelets undergo a release reaction and aggregate when they are exposed to aggregating substances

Table 5. Coagulation Abnormalities in Disease States

Condition	Laboratory Tests					
	PT	PTT	TT	Fib	FSP	Plt
Liver disease	−	−	−	∓	∓	∓
Vitamin K deficiency	−	−	+	+	+	+
DIC syndrome	−	−	∓	∓	∓	∓
von Willebrand's disease	+	−	+	−	+	+
Hemophilic A/B	+	−	+	+	+	+
Anticoagulant	∓	∓	∓	+	+	∓
Hypercoagulability	∓ (Short)	∓	∓	∓	∓	∓

+, normal; −, abnormal; PT, prothrombin time; PTT, partial thromboplastin time; TT, thrombin time; Fib, fibrinogen; FSP, fibrin(ogen) split products; Plt, platelet count.

such as collagen, adenosine diphosphate (ADP), epinephrine, and ristocetin. Abnormalities in platelet aggregation have been recognized in a variety of clinical disorders. Platelets from patients with von Willebrand's disease show a poor aggregation with ristocetin, whereas platelets from hemophilia A patients aggregate normally with ristocetin.

Clot retraction is used as an indication of platelet activity and platelet count and is influenced by a low concentration of fibrinogen in excessive fibrinolytic activity. Normal clot retraction begins within an hour and is complete within 24 hours. After complete clot retraction, the serum volume is approximately equal to the volume of the clot. Tests that are most commonly employed for clotting factor defects are prothrombin time (PT), partial thromboplastin time (PTT), fibrinogen, plasmin, fibrin monomers, and fibrin(ogen) split products. Prothrombin time determines plasma defects in the extrinsic system (Factor VII) as well as defects in Factors X, V, prothrombin, and fibrinogen. If the prothrombin time is prolonged and the fibrinogen level is adequate, the defect reflects either factor deficiency or inhibition of factors involved. The clotting time (PTT) measures clotting factors XII, XI, IX, VIII, X, II, and I and a prolonged PTT indicates factor deficiency or inhibition. Fibrinogen levels, plasmin activity, presence of fibrin monomers, and fibrin(ogen) degradation products are indicative of fibrinolytic processes and these parameters are investigated by functional and immunologic (mostly radioimmunoassay) methods. Table 5 lists functional parameters that are affected in various pathologic disorders.

Table 6. Peptide Markers of Diagnostic Importance

Level of Activation	Immunoassay
Coagulation	
Fibrinopeptide A	Radioimmunoassay (RIA)
Soluble fibrin monomers	Coaguloradioimmunometric assay
Fibrinolysis	
$B\beta_{15-42}$ peptide	RIA
Fibrin(ogen) degradation products (FDP)	RIA, latex agglutination
Fragment E antigen	RIA
Platelets	
Platelet factor 4 (PF4)	RIA
β-Thromboglobulin (BTG)	RIA
Thromboxane B_2 (TXB$_2$)	RIA
6-Keto-PGFI$_\alpha$	RIA
PGH$_2$ synthetase	IRMA
Leukotrienes	RIA

Conclusions

The physiology of hemostasis is regulated by a series of biochemical mechanisms which maintain the blood in its fluid state within the blood vessel. The various cellular and noncellular components which contribute to overall hemostatic functions of the body are endothelium, blood platelets, and the coagulation proteins. Any abnormality in absolute concentration or in biologic function of these components will result in hemostatic disorders. In a coagulation laboratory, a particular coagulation defect is evaluated by a number of biochemical, immunochemical, and radioisotopic methods. After establishing the cause and nature of coagulopathy, therapy is initiated and the progress of treatment is monitored by specific functional bioassays or immunologic methods. Some peptide markers of diagnostic importance which are determined exclusively by immunoassay are listed in Table 6.

Suggested Readings

Fareed J, Messmore, HL, Fenton JW, Brinkhous KM (eds): Perspectives in Hemostasis. New York, Pergamon Press, 1981

Thompson JM (ed): Blood Coagulation and Hemostasis, A Practical Guide. New York, Churchill Livingstone, 1980

3

Basic Concepts in Clinical Immunology

With the rapid advancements in immunochemistry and protein analytic techniques, our present day concept of immunologic terms and definitions has been greatly modified. Older definitions, although still in use, have been redefined as new information on the molecular properties of immunogenic proteins became available. In the following section, a brief account of modern immunologic terms is presented, so that the clinical application, interpretation, as well as limitations of each immunologic technique are understood in proper perspective.

Antigens (Immunogens)

Antigens are substances that induce formation of antibodies in animals and react with these antibodies. A more precise definition now recognizes that antigens are substances that can prompt B lymphocytes and T lymphocytes to produce specific responses to the antigen. Specific responses of B lymphocytes and T lymphocytes result in the production of antibodies and lymphokines that ultimately alter host cell behavior. Proteins, glycoproteins, and high molecular weight carbohydrates are most effective as antigens; purified fatty acids have not been shown to elicit effective antibody response.

Antigenic Determinant Sites

When an antigen is injected into an animal species, the immunologic response from the host is directed mostly toward a limited segment of the antigen, called the antigenic determinant site. Antigenic determinants consist of 3 or 4 amino acids located in one peptide chain plus

Table 7. Functional Properties of Major Immunoglobulins

Properties	IgG	IgM	IgE	IgA
Molecular weight	150,000	900,000	200,000	160,000
Electrophoretic mobility	γ	γ-B Fast	γ Fast	B
Concentration (mg/ml)	8–16	0.5–2	0.1–0.7	1.4–4.0
Complement fixation	+	+	–	–
Cross placenta	+	–	–	–

+, yes; –, no.

another group of 3 or 4 amino acids in another peptide chain. In general, there are two or more antigenic sites in one molecule, and these sites determine the specificity of the antigen–antibody reaction.

Antibodies

Antibodies are defined as proteins that are formed by the host animal as a reaction to a foreign antigen and can react specifically with it. The group of serum proteins to which antibodies belong is known as the immunoglobulins (Ig); these are specific to each determinant site in the antigen. Consequently, there are as many antibodies as there are antigenic determinants, and as a result, antibodies to identical antigens may be cross-reactive. In humans, mostly IgG (molecular weight [MW] 150,000) and IgM (MW 900,000) form the major classes of antibodies.

Antigen Binding Sites

The part of the antibody molecule which binds antigen consists of a small number of amino acids in the variable V region of the heavy (H) and light (L) chains. Some functional properties of major classes of immunoglobulins are presented in Table 7.

Structure

A typical antibody molecule (IgG) is composed of two H and two L polypeptide chains, held together by interchain disulfide bonds. N-terminal portions of both H and L chains show considerable variation (V region), whereas remaining parts of the chain are relatively constant in

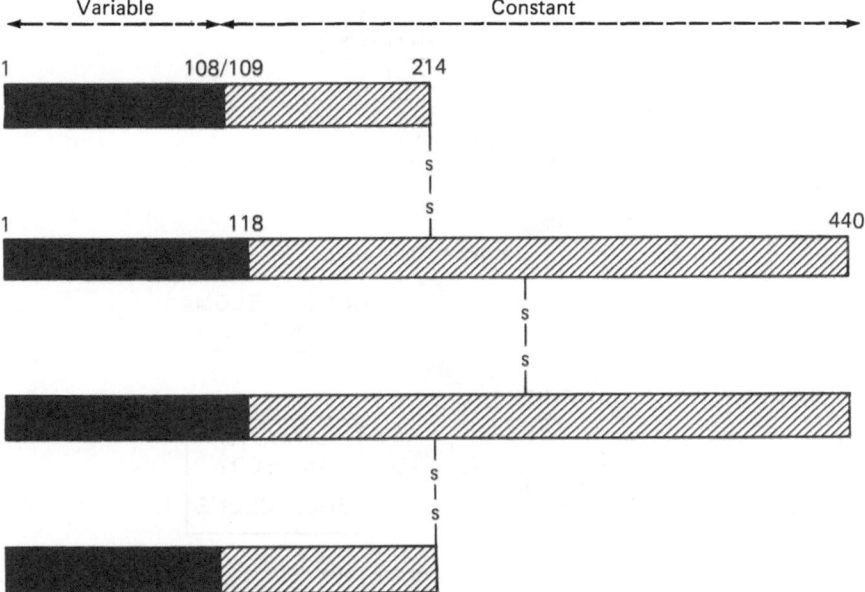

Figure 7. A diagramatic model of an IgG human antibody molecule, showing light and heavy chains.

composition. A representative structure of an immunoglobulin molecule is depicted in Fig. 7.

Antibody Specificity

As all antibodies are proteins, their specificity is defined by their amino acid sequence. The H and L chains of an antibody molecule consist of variable and constant regions and it is in these regions that antibody specificity is located (Fig. 7).

Monoclonal Antibodies

The classical method of antibody production often results in monospecific antiserum; however, such antisera often contain antibodies to different parts of the antigen, making them heterospecific in nature. Therefore, it is essential that cross-reacting antibodies be removed before the antiserum is utilized for a particular investigation. Koehler and Milstein[57] introduced a new concept of antibody production which is commonly referred to as hybridoma technology. Antibodies obtained from hybridoma technique are termed monoclonal antibodies and are both

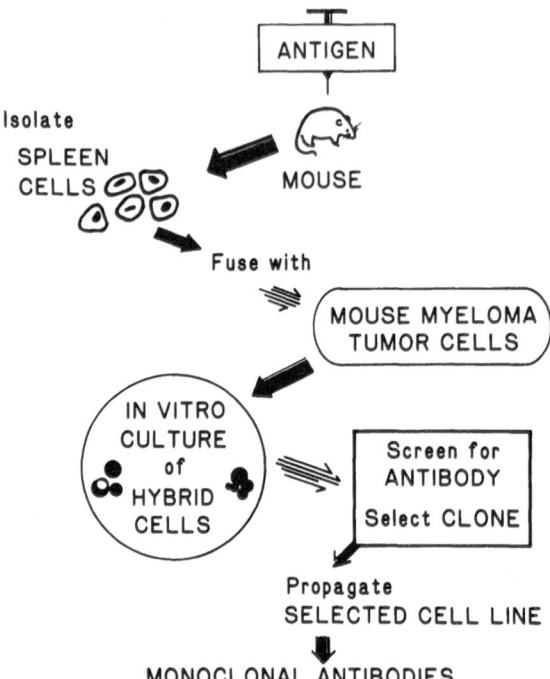

Figure 8. Stages in the production of monoclonal antibodies.

monospecific and monoclonal. Since these antibodies are secreted by a single hybrid clone their properties remain constant, and they eliminate the need to recalibrate the assay with each new batch of antiserum. A generalized scheme of hybridoma technology is depicted in Fig. 8.

Monoclonal antibodies to Factor VIII : RAg, Factor IX, Factor V, and α_2-antiplasmin have been developed; at present only the mouse mono-clonal antibody to Factor VIII : RAg and platelets (GP11b/111a com-plex) is commercially available from Cappel Laboratories, Westchester Pennsylvania. Hybridoma technique has been applied to characterize platelet glycoproteins and to purify Factor V from plasma by immu-noadsorption.

Suggested Readings

Fudenberg HH, Stites DP, Caldwell JI, Wells JV (eds): Basic and Clinical Immu-nology. Las Altos, Lang Medical, 1976.

Hurrell JGR (ed): Monoclonal Hybridoma Antibodies: Techniques and Applica-tions. Boca Raton, CRC, 1983.

4

Review of Immunologic Techniques

Immunologic methods provide highly specific and unique tools to determine minute amounts of antigenic proteins in serum, plasma, and other biologic fluids. Such methods of protein detection are called immunoassays. In general, immunoassays are classified as type I and type II immunoassays depending upon the quantitative use of antibody. Type I assays utilize an excess of antibody compared to the amount of antigen. In type II assays the amount of antibody is less than that of the antigen, and as such, these assays are also called saturation assays. The specificity of the type I system implicitly depends on purity of the antibody, whereas in type II, absolute homogeneity of antibody is less important, thereby making them more specific than type I assays since cross-reactants are less avidly inherently bound to the principal antibody. Of the various immunoassays, only radial immunodiffusion, rocket immunoelectrophoresis (electroimmunodiffusion; Laurell technique), and radioimmunoassays are frequently used to quantitate coagulation proteins. However, more sophisticated immunochemical methods such as nephelometric, fluorometric, and enzyme-linked immunosorbent techniques have also been introduced recently and some of these newer methodologies are readily available. A brief description of each immunoassay is given in order to elucidate the fundamental principles upon which these assays are based. The procedures are greatly simplified and are explained by graphic illustrations.

Radial Immunodiffusion (RID)

The principle of two-dimensional immunodiffusion was first described by Ouchterlony[81] and was later applied to quantitative determination of proteins. In this method, it is assumed that the amount of an antigen

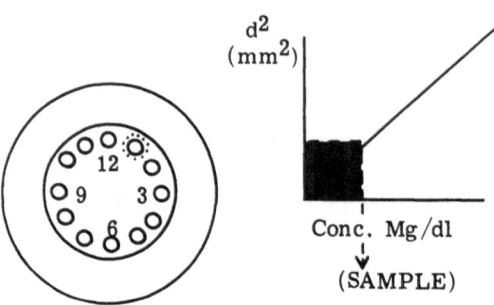

Figure 9. A schematic representation of radial immunodiffusion (RID) method.

placed in an antiserum-containing agarose bears a linear relationship to the size of the precipitate ring produced as a result of diffusion (Fig. 9). There are two modifications of this basic method. In the Mancini technique[73] the area of a given precipitate is allowed to reach a maximum size depending on the antigen concentration, and the area of the precipitin ring (d^2) is then linearly proportional to the antigen concentration. It is, therefore, essential that ring measurements are made after the precipitin ring reaches its maximum, otherwise d^2 will not bear a linear relationship with antigen concentration. In the Fahey and McKelvey[29] method, the precipitate rings are measured before their full development, and the diameter of the ring is taken to be proportional to the logarithm of the antigen concentration. Since the rate of precipitate formation is measured, it is absolutely important that the ring diameters of the standard and the unknown sample are read at the same time and at the same experimental conditions.

In most coagulation laboratories, RID is widely utilized because of its simple technique and reproducibility. Many commercially prepared RID plates utilize Mancini or Fahey–McKelvey principles and are available for measuring fibrinogen, plasminogen, antithrombin-III, α_1-antitrypsin, and α_2-macroglobulin. Ritzman[95] has discussed the relative advantages and limitations of this technique in quantitating serum proteins.

Electroimmunodiffusion (EID)

Rocket Immunoelectrophoresis

The technique of immunoelectrophoresis for the separation of proteins was first adopted by Grabar and Williams[40] and has remained a powerful analytic tool. In rocket technique the antigen is allowed to migrate under an electric field, through an antibody-containing medium, and the precipitation patterns which are formed in the shape of "rockets" are measured. The length of the rocket is proportional to the concentration of

ELECTRO
 IMMUNO
  DIFFUSION

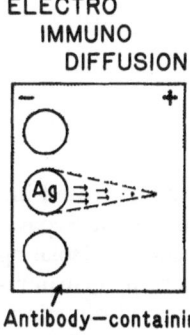

Antibody—containing
Agarose

Figure 10. Principles of electroimmunodiffusion (rocket method). Antigen–antibody interaction results in precipitin arc resembling a rocket shape.

antigen. The technique is commonly referred to as Laurell technique or rocket immunoelectrophoresis. Quantitation of proteins can be performed by single one- or two-dimensional electroimmunodiffusion. In single one-dimensional Laurell method (Fig. 10), antigen is allowed to move through agarose-containing specific antiserum. Precipitation occurs in the form of an ascending rocket. The height of the rocket is then measured and is proportional to the antigen concentration. Various factors which affect the performance of the assay are summarized below:

1. Antigen
   a. Purity
   b. Concentration
   c. Molecular weight
2. Antibody
   a. Origin
   b. Titer
   c. Specificity
3. Buffer
   a. pH
   b. Ionic strength
4. Agarose
   a. Purity
   b. Concentration
5. Electrophoresis
   a. Voltage
   b. Duration
   c. Temperature
6. Staining
   a. Concentration
   b. Duration
   c. Destaining

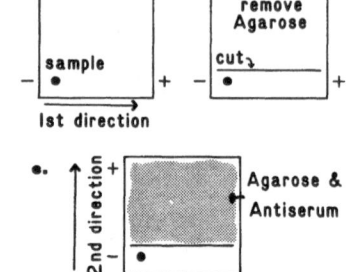

Crossed Immunoelectrophoresis

Figure 11. Stages in performing crossed immunoelectrophoresis of proteins. **a** Separation of proteins in first direction. **b** Isolating separated proteins. **c** Antiserum incorporation and electrophoresis in second direction.

Crossed Immunoelectrophoresis (CIEP)

First applied by Ressler[92] and later modified by Laurell,[63] CIEP is used to detect molecular abnormalities in proteins and is capable of separating proteins differing slightly in their isoelectric points. In this two-dimensional electroimmunodiffusion, the test sample is first subjected to electrophoresis in agarose. A longitudinal strip with separated proteins is cut, remaining agarose removed from the plate, and a mixture of agarose–antiserum is then poured on the exposed glass surface. The second electrophoresis is performed at 90° to the first run. A stepwise operation of this technique is shown in Fig. 11. A modified CIEP technique was described by Kuusi,[59] in which it is not necessary to remove the agarose after the first directional electrophoresis. A small piece of filter paper soaked in 1 ml of diluted antiserum is overlaid on the agarose surface and the electrophoresis in the second direction immediately follows. This procedure has been tried in our laboratories and was found to be a convenient and equally sensitive modification of Laurell's original method.

Electroimmunodiffusion, in general, offers several advantages over the time-consuming immunodiffusion methods. One can obtain results within a few hours as compared to 48–72 hours required in RID. In addition to free antigen, complexed forms such as plasmin : α_2-antiplasmin, plasmin : α_2-macroglobulin, thrombin : antithrombin-III, as well as molecular aberrations in clotting factors, can be identified on the basis of molecular configuration.

Turbidimetric Immunoassay (TIA)

The principle of turbidimetry rests on the observation that there is a linear relationship between optical density and the amount of precipitate in a solution. This technique was first applied to measurement of proteins in 1938, and since then several automated systems have been developed. Turbidity of an antigen–antibody complex is measured in a spectrophotometer and the sensitivity of the methods ranges between 0.5 and 1 μg/ml.

Recently, a turbidimetric immunoassay has been developed to measure plasma fibronectin, which will be discussed below. In addition to the manual methods, a highly automated, particle enhanced turbidimetric immunoassay for drugs* has been described by investigators at E.I. du Pont de Nemours & Company, Wilmington, DE. Similar automated immunoassays for Factor VIII : RAg and protein C are being developed.

Nephelometric Immunoassays (NIA)

The term nephelometry refers to the measurement of light scatter in a turbid solution. The principle of nephelometry was applied during the early part of this century by chemists and biochemists, and did not find its place in clinical immunology until recently. The first clinical application of immunonephelometry was attempted by Schultze and Schwick in 1959,[99] who quantitated several plasma proteins and used spectrophotometers to measure turbidity. Using this approach, several nephelometric methods have been utilized to measure proteins in serum, urine, cerebrospinal fluid, and other body fluids.[3,11,13,14,25,26,32,52–54,84,94,108,113] In earlier nephelometric assays, the intensity of light scatter was measured at 90° to the angle of incident light, which also amplified the background scatter signal. In order to eliminate background scatter and to obtain increased sensitivity, several improvements have been incorporated in commercial nephelometers. Most of these nephelometers utilize a laser beam as the light source and detect only the forward light scatter. Background scatter from dust and contaminating particles is reduced significantly by an electronic device. The reagent system includes a polymeric buffer, which enhances the insolubility of antigen–antibody complexes and increases the intensity of light scatter and its detection. Laser light is used because of its following characteristics:

1. High intensity—nearly all energy from the laser goes into the narrow beam.
2. High degree of collimation—it can be focused into a thin beam.
3. Monochromatic—the light is of only one color; thus it is ideal for light-scattering techniques.

---

* Cited by Litchfield WJ, et al: High sensitivity particle enhanced turbidimetric immunoassays. Abstract #3742. Presented at the 67th Annual Meeting of Federation of American Societies of Experimental Biology, April 10–15, 1983, Chicago.

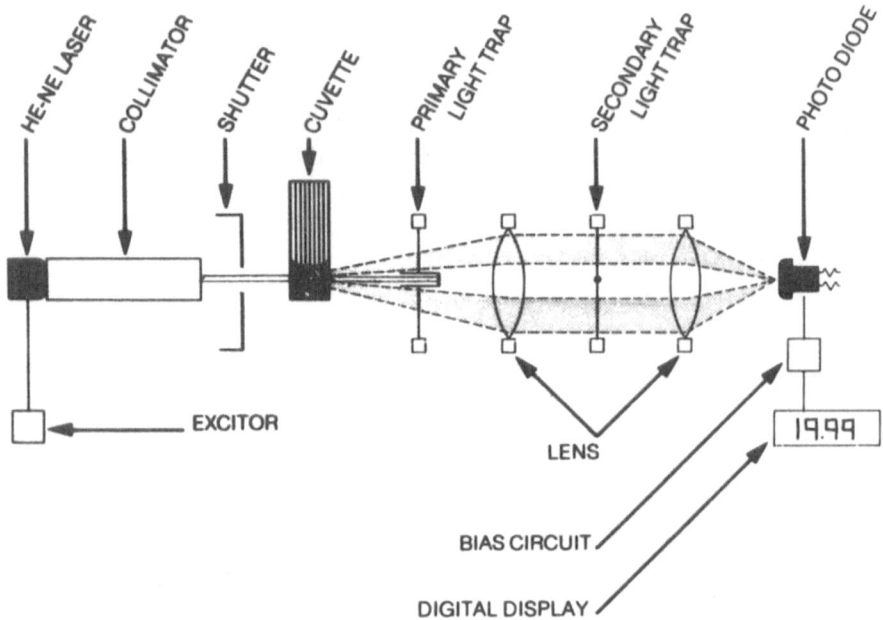

Figure 12. Measurement of light scatter in an antigen–antibody solution. Courtesy of Calbiochem Behring Corp.

A graphic representation of its operating principles is given in Fig. 12.

In an end-point nephelometric immunoassay, an incubation time ranging from 1 to 2 hours is generally required, whereas in rate nephelometric procedures, the rate of antigen–antibody complex formation is measured, and therefore no incubation time is needed. In a kinetic mode laser nephelometer immunology series 420 (Fig. 13), the rate of change in light scatter is monitored by taking a light scatter reading every 50 milliseconds during the 24-second reaction period. The ratios of the reference and sample readings are evaluated by the microprocessor. In order to eliminate the noise signal, spikes in the scatter readings are eliminated, and remaining readings are averaged for every 250 milliseconds. This yields four data points per second for a total of 96 data points. When the last data point is processed peak rate is calculated. Sample concentrations are calculated from a reference curve generated by the assay of six increasing concentrations of reference material. A cubic curve is fitted through a plot of peak rate vs. assigned concentration.

In actual procedure, a solution containing antigen (e.g., AT-III in plasma) is mixed with its monospecific antiserum in a cuvette, and after a suitable incubation time the cuvettes are placed in a nephelometer. A laser beam is then passed through the solution and the amount of light

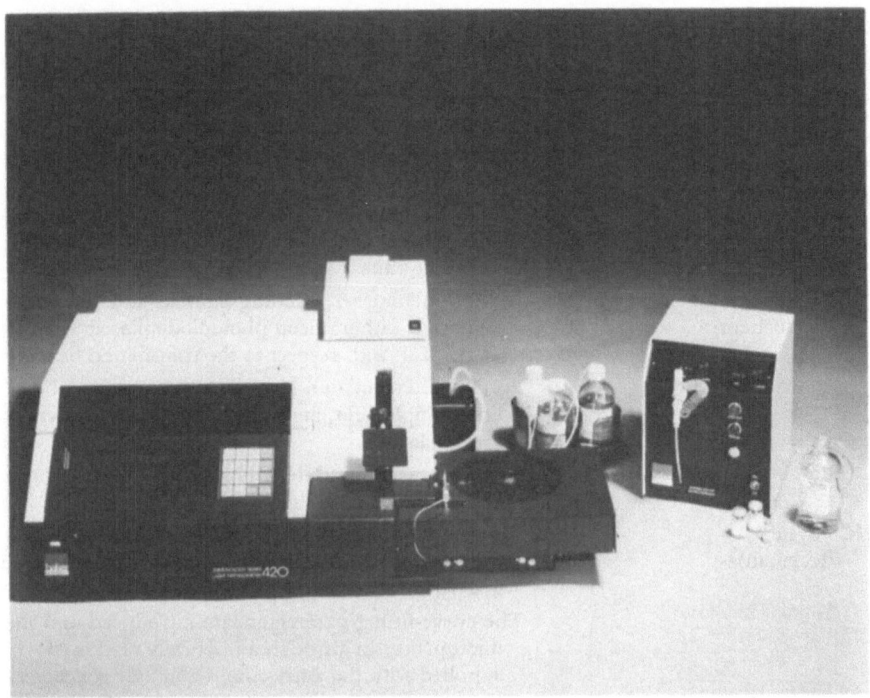

Figure 13. Laser Nephelometer Immunology Series 420. Courtesy of Baker Instruments.

scattered by the antigen–antibody complexes is measured. Nephelometric methods are now increasingly used to determine AT-III, α_1-AT, and Factor VIII : RAg, and are reported to be reproducible, highly sensitive, and bear a good correlation with RID method. A list of commercially available nephelometers is given in Table 8. Because of their nonisotopic nature and the potential for complete automation, it is likely that rate and end-point nephelometric assays will play a key role in the rapid quantitation of coagulation proteins. Plasma proteins quantitated by various nephelometers are shown in Table 9.

Suggested Readings

Huebner SO, Nachbar J, Asbeck F: The determination of antithrombin-III, α_2-macroglobulin, α_2-antiplasmin in plasma by laser nephelometry. Clin Chem Clin Biochem 18:221–225, 1980.

Parvez Z, Fareed J, Messmore HL, Moncada R: Laser nephelometric quantitation of antithrombin III (AT-III). Development of a new assay. Thromb Res 24:367–377, 1981.

Table 8. Comparative Features of Commercial Nephelometers

Nephelometer	Features
Laser Nephelometer PDQ (Hyland)	Helium–neon laser, emitting photons at 632.8 nm; equipped with a photomultiplier tube mounted at an angle of 31° to the incident light.
	Individual LAS-R test kits are available for immunoglobulins, protease inhibitors, complement, and other acute phase proteins.
	The microprocessor module, which is optional, calculates protein concentration in conventional as well as WHO units.
Laser Nephelometer System (Calbiochem)	Helium–neon laser, emitting photons at 632.8 nm; equipped with a silicon photodiode placed at an angle of 0° with respect to the transmitted beam.
	Protein determination kits for most immunoglobulins, complement, and other acute phase proteins are available.
	Data processing module (DPM) prints out concentration in mg/dl or IU/ml.
ICS System (Beckman)	Tungsten/iodine lamp, emitting photons at 400–550 nm; equipped with a detection optic system located at a 70° forward angle.
	The curve-fitting parameters are introduced into the microprocessor through an optically read card supplied with the antiserum. Calibrator concentrations are read in from calibrator-read card.
	The results are displayed on an alpha-numeric display in mg/liter, mg/dl, or IU/ml.
	Protein determination kits for immunoglobulins, albumin, complement, and other proteins are available.
Immunology Series 420 (Baker)	Polarized helium–neon laser emitting photons at 632.8 nm; angle of measurement is 37.8°.
	Throughput rate is 25–30 samples/hour with a precision of ±5% coefficient of variation.
	Ten protocols are stored in the microprocessor which generates 6-point reference curve automatically. Alphanumeric printer provides a hard copy printout of concentration values.
	Serum protein assays are available for most immunoglobulins, haptoglobin, α_1-antitrypsin, α_2-macroglobulin, C3, C4, and albumin.
Multistat III; F/LS (Instrumentation Lab)	Centrifugal analyzer with both fluorescence/light scatter and conventional absorbence optical paths.
	Completely automated loading, analysis, and data acquisition. Sample and reagent volumes are 10 μl and 150 μl, respectively.
	Nephelometric assays for α_1-antitrypsin and antithrombin-III have been adapted and will be commercially available soon.

Table 9. List of Plasma Proteins Quantitated by Nephelometric Immunoassay

Instrument	Proteins
Laser Nephelometer PDQ (Hyland)	Human IgGA; human IgG/CSF, IgG; human IgM; human complement C3; human CSF albumin; α_2-macroglobulin; ceruloplasmin; C3 proactivator; C1 esterase inhibitor; α_1-acid glycoprotein; plasminogen; prealbumin; human α_1-antitrypsin; human complement C4; rheumatoid factor; C-reactive protein
Laser Nephelometer System (Calbiochem)	IgG; IgM; IgA; α_1-antitrypsin; complement component C3; C4; C3 proactivator; albumin; haptoglobin; ceruloplasmin; transferrin; β-lipoprotein
Immunochemistry System (Beckman)	α_1-Antitrypsin; C3 proactivator; ceruloplasmin; α_1-acid glycoprotein; α_2-macroglobulin; C-reactive protein; IgG; IgM; IgA; C3 complement; C4 complement; albumin; transferrin
Immunology Series 420 (Baker)	IgG; IgM; IgA; complement C3; complement C4; hemopexin; C-reactive protein; albumin; α_2-macroglobulin; haptoglobin

Radioimmunoassay (RIA)

The principle of radioimmunoassay was first applied by Yalow and Berson[119] in 1960 to measure the endogenous insulin in human plasma. Since then, the method has been extensively utilized in many routine clinical diagnostic tests. However, RIA techniques were not fully used in the assessment of coagulation disorders until the late 1970s. It was only after 1974 that radioassays were employed to study the molecular configuration of fibrinogen, prothrombin, and other vitamin K–dependent factors, and have since then contributed greatly to our understanding of the mechanisms of thrombosis and its diagnosis.

RIA procedure is based on the competitive binding of antigen to a monospecific antibody. The unlabeled antigen competes with radiolabeled antigen for the antibody. Upon addition of unlabeled antigen, the amount of free (unbound) radiolabeled antigen is proportional to the quantity of unlabeled antigen added. A standard curve is constructed by plotting the percentage of antibody-bound radiolabeled antigen against the known concentrations of a standard unlabeled antigen. A typical radioimmunoassay procedure is presented in Fig. 14. As can be seen, there are four basic components in a radioimmunoassay:

1. A highly purified antigen sample.
2. A monospecific antiserum.
3. Radiolabeled purified antigen.
4. A suitable method for the separation of antigen–antibody complex from free antigen.

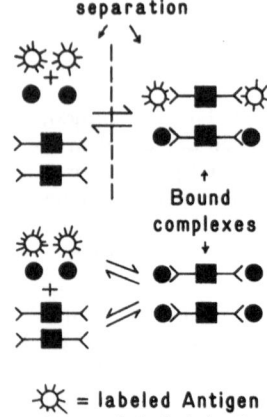

☼ = labeled Antigen
● = unlabeled Antigen
>─■─< = Antibody

Figure 14. Competitive binding of an unlabeled antigen to its monospecific antiserum in a radioimmunoassay.

Although RIA is by far the most sensitive method for the detection of antigens or antibodies,[72,79] it requires expensive equipment, highly purified reagents, and separation procedures. In coagulation testing, radioimmunoassays are employed to quantitate Factor VIII : RAg, Factor VIII : CAg, AT-III, platelet factor 4, βTG, fibrinopeptide A, and Bβ15-42 peptide, as these proteins are increasingly utilized as molecular markers of hemostatic pathway defects.

Immunoradiometric Assay (IRMA)

The immunoradiometric assay was developed for measuring polypeptide hormones and since then has been applied to detect several other proteins.[34,37,62,68,78,86] IRMA differs from RIA in that antibodies instead of antigen are iodinated. The procedure (Fig. 15) consists of adsorbing

←O→ Antigen
>─□─< Antibody
>─■─< I^{125} —labeled antibody

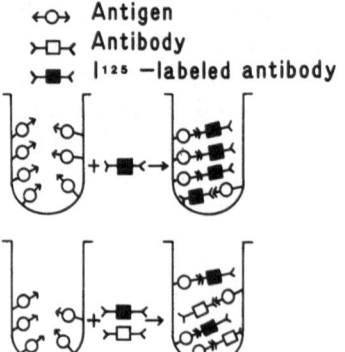

Figure 15. Competitive binding of nonlabeled antibody to a fixed antigen in an immunoradiometric (IRMA) assay.

an antigen to a solid surface (plastic tube) and then adding radiolabeled antibodies to saturate the fixed antigen. Serum samples containing antibodies to the antigen are added to the test tube. The nonlabeled antibodies from the sample will compete with the labeled antibodies for the antigen and some of the labeled antibodies will remain free. The percentage of free labeled antibodies not bound to the antigen is related stoichiometrically to the nonlabeled antibodies in the sample.

IRMA methodologies have been successfully applied by European researchers to quantitate Factor VIII : CAg, Factor VIII : RAg, fibrino-peptide A, Bβ15-42, and antithrombin-III. As nonisotopic methods are not yet available for measuring these proteins, the use of IRMA and RIA will be continued for some time.

Suggested Readings

Felber JP: Radioimmunoassay in the clinical chemistry laboratory. Adv Clin Chem 20:130, 1978.

Peak IR, Bloom AL: The use of an immunoradiometric assay for Factor VIII–related antigen in the study of atypical von Willebrand's disease. Thromb Res 10:27–32, 1977.

Fluoroimmunoassay (FIA)

Fluorescence is the emission of light of one color, i.e., wavelength, when a substance is irradiated with light of a different color. Fluorochromes such as rhodamine or fluorescein have characteristic absorption and emission spectra. Fluorescein isothiocyanate (FITC) has an absorption maximum at 490–495 nm and shows a maximal emission (green color) at 517 nm. The technique of immunofluorescence was introduced by Coons in 1941 who employed β-anthracene, a blue fluorescent compound, coupled to pneumococcus antiserum to detect bacterial antigens in tissue sections. In principle, FIA measures the amount of fluorescent antibody consumed in an immunochemical reaction.[20,39,67,120] The immunochemical reaction takes place in two phases (Fig. 16). In the liquid phase, the antigen in the sample is allowed to react with excess of mono-specific fluorescent antibody against the antigen. The solid phase of the assay consists of immunoadsorption of unreacted fluorescent antibody by the antigen previously coated on the surface of a solid matrix. Fluorescence on the matrix is then measured in a fluorometer, and is inversely proportional to the antigen concentration in the sample. FIA is not currently available for measuring coagulation proteins. However, these assays have been used in research and were shown to correlate well

Liquid phase reaction takes place in tube

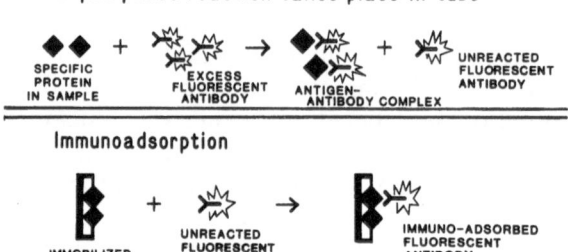

Immunoadsorption

Figure 16. Solid- and liquid-phase stages in a fluoroimmunoassay for Factor VIII : RAg in plasma.

with other methods. The technique has been applied to measure Factor VIII : RAg, α_2-antiplasmin, and fibrinopeptide A.

Although fluoroimmunoassay resembles RIA in principle, FIA is considerably safer and does not require radioiodination procedures and expensive equipment.

Suggested Readings

Parvez Z, Fareed J, Messmore HL, Bermes EW: A rapid fluoroimmunoassay for the quantitation of Factor VIII : RAg in human plasma. In Sunderman W (ed): Manual of Procedures for the Laboratory Diagnosis of Disorders of the Fetus, Newborn and Infant. Philadelphia, Institute for Clinical Science, 1981, pp 101–110.

Rudzki Z, Tunbridge LJ, Lloyd JV: A new simple assay for Factor VIII–related antigen. Thromb Res 16:577–586, 1979.

Enzyme-linked Immunosorbent Assay (ELISA)

The principle of ELISA is similar to that of FIA with the exception that an enzyme–antibody conjugate is utilized instead of the fluorescein-labeled antibody. Although several modifications of ELISA techniques have been reported, in a simple assay (Fig. 17) a purified antigen is adsorbed onto a solid phase such as polypropylene test tube and the unadsorbed antigen is then washed with a buffer. Plasma containing the antigen is incubated with its monospecific antiserum raised in rabbits and is added to the tube coated with purified antigen. The unbound antigen from this mixture binds to the antigen on the tube surface and any nonreacted protein is washed with buffer. For the third step, enzyme is conjugated to a second antibody against rabbit IgG raised in goat

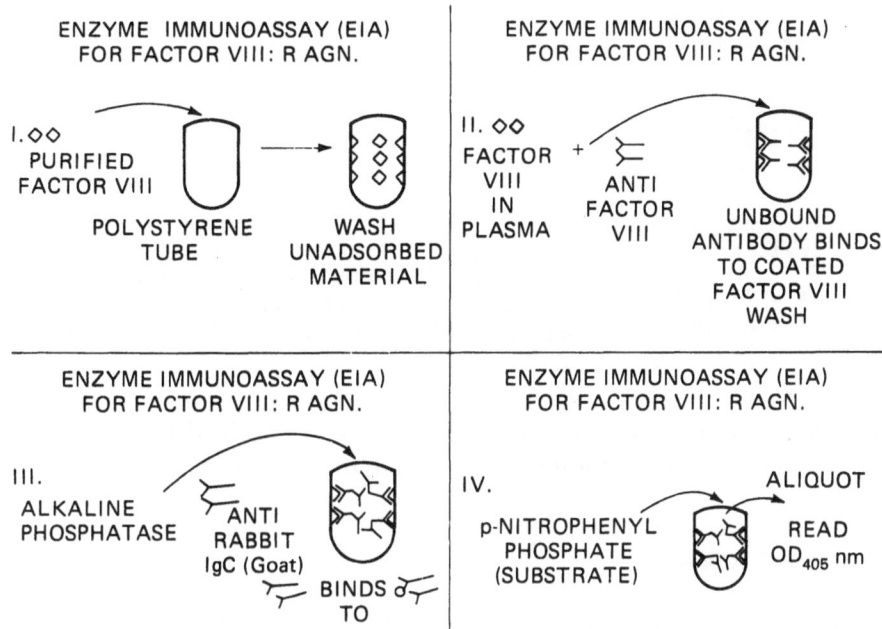

Figure 17. An ELISA procedure for Factor VIII : RAg in plasma. I, adsorption of purified antigen; II, incubation of antigen and antibody and separation of unbound antigen; III, conjugation of alkaline phosphatase to anti–rabbit IgG, which binds to anti–Factor VIII; IV, hydrolysis of the substrate by unbound-enzyme conjugate.

and this enzyme–antibody conjugate is added to the tube. The second antibody binds to the first antibody as it is recognized as an antigen. An enzyme-specific substrate is then added to the tube and an aliquot of this mixture is tested for hydrolysis of the substrate by the remaining enzyme, by monitoring the change in optical density at 405 nm in a spectophotometer. The degree of substrate hydrolysis is inversely related to the amount of antigen present in the sample.

In recent years, several EIA and ELISA procedures have been developed for antithrombin-III,[107,109,110] α_2-antiplasmin, Factor VIII : RAg, fibrinopeptide A, and antibodies to Factor IX; however, these methods are at an experimental stage and further studies are needed to establish their clinical usefulness. Although ELISA methodologies are much safer than the radioimmunoassays, endogenous enzymes in the plasma may interfere with the enzyme employed for labeling and an additional chemical step of measuring enzyme activity is required. In addition to this extra step, changes in temperature influence enzymatic activity, thus increasing the imprecision of the assay. Because of their high sensitivity and similarity to the radioimmunoassay, ELISA techniques can be read-a)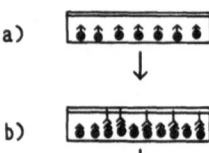

b)

c)

ↆ = Antigen
⅄ = Antibody
∞∞ = Water vapor

Figure 18. Operational steps in a thin-layer immunoassay. a, adsorption of antigen to the solid surface; b, addition of antibody; c, visualization of antigen–antibody reaction.

ily adapted to such automated instruments as the Gilford PR-50, Quantum-II, and other computer-assisted analyzers.

Suggested Readings

Dito WR: Nonisotopic immunoassay. Diagn Med 4:11–18, 1981.
O'Beirne AJ, Cooper HR: Heterogenous enzyme immunoassay. Histochem Cytochem 27(8):1148–1162, 1979.

Thin Layer Immunoassay (TIA)

Thin layer immunoassay (Fig. 18) is a solid-phase immunoadsorbent technique which resembles the ELISA procedure. In this assay a flat plastic surface is spotted with an antigen which is adsorbed on the surface. After washing the unadsorbed materials with buffer, a monospecific antibody is added to the spots and the reaction is allowed to proceed to completion. Residual material is washed off. Antigen–antibody complexes are visualized by exposure of the plate to water vapor at 60°C for 1 minute, and the condensation patterns are photographed. Antigen–antibody reaction can be recognized as hydrophilic areas with large condensation drops.

TIA has not been applied to quantitation of coagulation proteins; however, this technique is used to determine carcinoembryonic antigen (CEA) and hepatitis B surface antigen (HB$_s$Ag).

Particle Counting Immunoassay (PACIA)

Particle counting immunoassay is similar to the latex agglutination method, except that nonagglutinated latex particles (0.80 μm) are

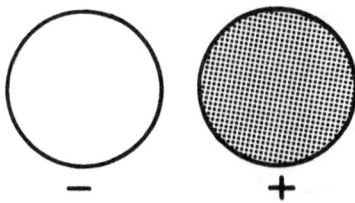

Figure 19. Detection of fibrin(ogen) degradation products in serum; agglutination is seen in the right (+) well.

counted. The latex particles are coated with a specific antibody, and upon addition of the antigen, agglutination of particles occurs. The method lends itself to automation and blood cell autocounters can be utilized. The procedure is simple, does not require radioisotopes or separation steps, and sensitivity of the method is reported to be as good as with an RIA method.

Latex Agglutination Test

The cross-linking of multivalent antigens by antibody results in a precipitation reaction, whereas cross-linking of cells or large particles by antibody directed against surface antigens leads to agglutination. Generally, agglutination reactions are used to identify bacteria and to type red cells; nevertheless, the same principle is utilized to detect soluble antigen by coating latex particles with specific antibody.

The latex agglutination or hemagglutination test has been recently introduced for a qualitative/semiquantitative determination of fibrin (ogen) degradation products in plasma, serum, and urine. The results can be best expressed as positive or negative (Fig. 19). The technique can detect both the late degradation products D and E as well as the early degradation products X and Y, and correlates well with the hemagglutination assay.

Clinical Considerations

In the past, coagulation disorders were studied mainly by quantitating concentration levels of a given protein and/or its biologic function. As a result, coagulation defects due to molecular aberrations in coagulation factors could not be identified either with functional assays or by quantitative immunoassays. However, molecular variant forms of fibrinogen, prothrombin, Factor VIII, Factor IX, and Factor X have been recognized by utilizing immuno- and crossed immunoelectrophoresis techniques. Neoantigens arising as a result of enzyme–inhibitor complex formation (thrombin : antithrombin, plasmin : α_2-antiplasmin, etc.) are often indicative of consumptive coagulopathies and these complexes can

Table 10. Relative Sensitivity of Immunologic Methods

Method	Detection Limit
Immunoprecipitation	10–20 μg/ml
Double diffusion	10–20 μg/ml
Radial immunodiffusion	10–20 μg/ml
Electroimmunodiffusion	5–20 μg/ml
Nephelometric	10–20 μg/ml
Radioimmunoassay	0.0005–0.005 μg/ml
Fluoroimmunoassay	0.00005–0.005 μg/ml
Enzyme immunoassay	0.00005–0.005 μg/ml
Latex agglutination	1–2 μg/ml
Hemagglutination inhibition	0.005–0.01 μg/ml

be detected by sensitive ELISA or RIA methods. These techniques are now used to assess molecular complexes of thrombin: AT-III, plasmin: α_2-AP, and plasmin: α_2-M in disseminated intravascular coagulation and other coagulation abnormalities in leukemia and cancer patients. However, results from individual methods should be evaluated in the light of other functional tests for a definitive diagnosis of any coagulation disorder. The relative sensitivity of each method adopted (Table 10) should also be considered. In plasma fibrinogen concentration ranges from 200 to 400 mg/dl and AT-III from 10 to 30 mg/dl, whereas platelet specific proteins PF_4 and β-TG fluctuate in ng/ml range. Depending upon the concentration of the protein, a suitable method should be employed. For example, immunoprecipitin assays are insensitive to low concentration of antigen, and hence PF_4, BTG, FPA, or $B\beta$15-42 peptides cannot be measured by single radial immunodiffusion or rocket technique. Success in immunodiagnosis of coagulation disorders will depend largely on careful choice of the method utilized.

Selected Bibliography

General

Allen PG: Plasma Proteins: Analytical and Preparative Techniques. Philadelphia, Blackwell Scientific, 1977.

Aloisi RM: Principles of Immunodiagnostics. St. Louis, Mosby, 1979.

Cawley LP: Electrophoresis and Immunoelectrophoresis. Boston, Little, Brown, 1969.

Clausen J: Immunochemical techniques for the identification and estimation of macromolecules. In Work TS, Work E (eds): Laboratory Techniques in Biochemistry and Molecular Biology. Amsterdam, North-Holland, 1977.

Crowle AJ: Immunodiffusion. New York, Academic Press, 1973.

Fudenberg LL, Stites DP, et al: Basic and Clinical Immunology. Los Altos, Lange Medical, 1976, pp 281–294.

Haywood BJ: Electrophoresis: Technical Applications. Ann Arbor, Science Publications, 1969.

Khristov V, Apostolov P, Stamenova T: Fibrin degradation products in the urine in diabetic nephropathy. Vutr Boles 18(4):32–36, 1979.

Kwapinski JBG: Methodology of Immunochemical and Immunological Research. New York, Wiley, 1972.

Laurell CB: Quantitative estimation of proteins by electrophoresis in agarose gel containing antibodies. Ann Biochem 15:45–52, 1966.

Marer RJ, Walker JH (eds): Immunochemical methods in the biological sciences: enzymes and proteins. London, Academic Press, 1980.

Martin JP, Lemercier JP: Immunological methods used for the diagnosis of alpha 1-antitrypsin deficiency. Bronchopneumologie 29(4):318–322, 1979.

Milgorm F, Abeyounis CJ, Kano K (eds): Principles of Immunological Diagnosis in Medicine. Philadelphia, Lea & Febiger, 1981.

Mittman C: Pulmonary Emphysema and Proteolysis. New York, Academic Press, 1972.

Ouchterlony O: Handbook of Immunodiffusion and Immunoelectrophoresis. Ann Arbor, Science Publications, 1968.

Richter MA: Clinical Immunology. A Physician's Guide. Baltimore, Williams & Wilkins, 1982.

Ritzman SE, Daniels JC (eds): Serum Protein Abnormalities—Diagnostic and Clinical Aspects. Boston, Little, Brown, 1975.

Roitt IM: Essential Immunology. Oxford, Blackwell Scientific, 1971.

Rose NR, Friedman H (eds): Manual of Clinical Immunology. Washington, D.C., American Society of Microbiology, 1976.

Rose NR, Milgrom F, Oss CJF: Principles of Immunology. New York, Macmillan, 1979.

Sugiura M, Akatsuka M, Hirano K, Saito Y: Comparison of biological and immunological methods for determination of serum alpha-1-antitrypsin. Chem Pharm Bull (Tokyo) 26(1):220–225, 1978.

Thompson RA: The Practice of Clinical Immunology. Baltimore, Williams & Wilkins, 1974.

Thompson RA (ed): Techniques in Chemical Immunology. Oxford, Blackwell Scientific, 1981.

Thorell JI, Larson SM: Radioimmunoassay and Related Techniques. Methodology and Clinical Applications. St. Louis, Mosby, 1978.

Voller A, Bartlett A, Bidwell D (eds): Immunoassays for the 80's. Lancaster, MTP Press, 1981.

Vyas GN, Stikes DP, Brechner G (eds): Laboratory Diagnosis of Immunologic Disorders. New York, Grune & Stratton, 1975.

Weir DM: Handbook of Experimental Immunology. Oxford, Blackwell Scientific, 1973.

Radial Immunodiffusion

Fahey JL, McKelvey EM: Quantitative determination of serum immunoglobulins in antibody–agar plates. J Immunol 94:84–90, 1965.

Ginsberg MH, Hoskins R, Sigrist P, Psinter RG: Purification of a heparin-neutralizing protein from rabbit platelets and its homology with human platelet factor 4. J Biol Chem 254(24):123–171, 1979.

Hollander W, Colombo MA, Kirkpatrick B, Paddock J: Soluble proteins in the human atherosclerotic plaque, with special reference to immunoglobulins, C_3-complement component, alpha-I-antitrypsin and alpha-2-macroglobulin. Atherosclerosis 34:391–405, 1979.

Mancini G, Carbonara AO, Hermans JF: Immunochemical quantitation of antigens by single radial immunodiffusion. Immunochemistry 2:235–245, 1965.

Ojutiku OO, Ladipo GO: Alpha-I-antitrypsin levels and phenotypes in some healthy Nigerians—a preliminary report. Nigerian Med J 8(6):511–513, 1978.

Oudin J: Methode di analyse immunochimique par precipitation specifique en milieu gelife. Acad Sci CR Seances 222:115–116, 1946.

Ritzman SE: Radial immunodiffusion revisited. I. Lab Med 9(7):23–33, 1978.

Rocket Immunoelectrophoresis

Adamson J, Mathiesen M: Rocket technique for determination of alpha-l-globulin antitrypsin. In Mittman C (ed): Pulmonary Emphysema and Proteolysis. New York, Academic Press, 1972.

Anonymous: Gel electrophoresis: Diagram and methods. Nature 227:681, 1970.

Axelsen NH, Kroll J, Week B (eds): A manual of quantitative immunoelectrophoresis, methods and applications. Scand J Immunol 2(Suppl 1):1, 1973.

Gandolfo GM, Torresi MV: The influence of heparin on the immunochemical evaluation of antithrombin-III by electroimmunoassay. Haematologica 63(5):512–520, 1978.

Gil CW, Fischer CL, Holleman CL: Rapid method for protein quantitation by electroimmunodiffusion. Clin Chem 17:501–504, 1971.

Kroll J: Quantitation of protein by electrophoresis in a cellulose acetate membrane impregnated with antiserum. Scand J Clin Lab Invest 21:187–189, 1968.

Kueppers F: Antigen–antibody crossed electrophoresis for typing of alpha-1-antitrypsin. In Mittman C (ed): Pulmonary Emphysema and Proteolysis. New York, Academic Press, 1972.

Laurell CB: Quantitative estimation of proteins by electrophoresis in agarose gel containing antibodies. Ann Biochem 15:45–52, 1966.

Merrill DA, Hartley TF, Claman HN: Electroimmunodiffusion (EID): a simple, rapid method for quantitation of immunoglobulins in diluted biological fluids. J Lab Clin Med 69:151–159, 1967.

Ressler RN: Electrophoresis of serum protein antigens in an antibody containing buffer. Clin Chim Acta 5:359–365, 1960.

Verbruggen R: Quantitative immunoelectrophoretic methods. A literature survey. Clin Chem 21(11):5–43, 1975.

Crossed Immunoelectrophoresis

Anderson LO, Engman L, Henningsson E: Crossed immunoelectrophoresis as applied to studies on complex formation. The binding of heparin to anti-thrombin-III and the antithrombin-III–thrombin complex. J Immunol Method 14:271–281, 1977.

Chan V, Chan TK: Heparin–antithrombin-III binding. In vitro and in vivo studies. Haemostasis 8(6):373–389, 1979.

Henriksson P, Milsson IM: Effects of leukocytes, plasmin, and thrombin on clotting factors. A comparative in vitro study. Thromb Res 16(3–4):301–312, 1979.

Johansson L, Hedner U, Milsson IM: Familial antithrombin-III deficiency as pathogenesis of deep venous thrombosis. Acta Med Scand 204(6):491–495, 1978.

Kelly JK, Taylor TV, Milford-Ward A: Alpha-1-antitrypsin PI$_s$ phenotype and liver cell inclusion bodies in alcoholic hepatitis. J Clin Pathol 32(7):706–709, 1979.

Kuusi N: A technical improvement for crossed immunoelectrophoresis. J Immunol Methods 31:361–364, 1979.

Sutltan Y, Jeanneau C, Lamaziere J, Maisonneuve P, Caen JP: Platelet factor VIII–related antigen. Studies in vivo after transfusion in patients with von Willebrand's disease. Blood 51(4):751–761, 1978.

Zimmerman TS, Abildgaard CF, Meyer D: The factor VIII abnormality in severe von Willebrand's disease. N Engl J Med 301(24):1307–1310, 1979.

Laser Nephelometry

Alper CA: Automated nephelometric determination of serum heptoglobulin, C_3 and alpha-1-antitrypsin. In Barten CE, et al. (eds): Advances in Automated Analysis. Miami, Thurman Association, 1971.

Buffone GJ, Savory V, Cross RE: Evaluation of kinetic light scattering as an approach to the measurement of specific proteins with the centrifugal analyzer. I. Methodology. Clin Chem 21(2):1731–1734, 1975.

Buffone GJ, Savory J, Cross RE: Use of a laser-equipped centrifugal analyzer for kinetic measurement of serum IgG. Clin Chem 20(10):1320–1323, 1974.

Deaton CA, Maxwell KW, Smith RS, Crevling RL: Use of laser nephelometry in the measurement of serum proteins. Clin Chem 22(9):1465–1471, 1976.

Dito WR: Rapid immunonephelometric quantitation of eleven serum proteins by centrifugal fast analyzer. Am J Clin Pathol 71(3):301–308, 1979.

Dunikoski LK, Myrmel KH, Derzack MT: Automated chromogenic antithrombin-III assay with a centrifugal analyzer. Clin Chem 25(6):1076, 1979.

Eckman I, Robbins JB, Vandenhamer CJA, Lentz J, Scheinberg IH: Automation of a quantitative immunochemical microanalysis of human serum transferrin: a model system. Clin Chem 16:58, 1970.

Giddings JC, Evans DJ, Bloom AL: Quantitation of factor VIII–related antigen (F VIII RAG) using a laser nephelometer. Thromb Res 15(5–6):847–855, 1979.

Kierulf P, Brosstad F, Godal HC, Lund PK, Anderson AB, Naeverlid I: Poor discriminating power of fibrinogen–sepharose toward plasma fibrinogen, fibrin des-aabb monomers, as studied by labeled proteins, laser nephelometric FR-antigen and RIA-FPA quantitation. Thromb Res 16(1–2):1–10, 1979.

Killingsworth LM, Savory J: Nephelometric studies of the precipitin reaction: a model system for specific protein measurements. Clin Chem 19:403–407, 1973.

Ritchie RF, Alper CA, Graves J, Pearson N, Larson C: Automated quantitation of proteins in serum and other biologic fluids. Am J Clin Pathol 59:151–159, 1973.

Romette J, Levy G, Dicostanzo J, Gar Boudouresques G: Determination of proteins in cerebrospinal fluid of adults using two immunochemical methods. Clin Chim Acta 94(2):121–124, 1979.

Savory J, Killingsworth LM: Determination of alpha-1-antitrypsin by a nephelometric procedure. Ann Clin Lab Sci 3:43–47, 1973.

Schultze HE, Schwick G: Quantitative immunologische bestimmung von plasma protein. Clin Chim Acta 4:15–26, 1959.

Sieber A, Bross J: Protein determination by laser nephelometry. Med Lab 2:17–24, 1977.

Sternberg JC: A rate nephelometer for measuring specific proteins by immunoprecipitin reactions. Clin Chem 23(8):1446–1456, 1977.

Tiffany TO, Parella JM, Johnson WF, Burtis CA: Specific protein analysis by light scatter measurement with a miniature centrifugal fast analyzer. Clin Chem 20(8):1055–1061, 1974.

Radioimmunoassay

Astedt B: No cross reaction between circulating plasminogen activator and urokinase. Thromb Res 14(4–5):535–539, 1979.

Brown JE, Baugh RD, Hougie C: Effect of exercise on the factor VIII complex: a correlation of the von Willebrand antigen and factor VIII coagulant antigen increase. Thromb Res 15(1–2):61–67, 1979.

Chan V, Chan TK: Heparin–antithrombin-III binding. In vitro and in vivo studies. Haemostasis 8(6):373–389, 1979.

Chan V, Chan TK, Wong V, Tso SC, Todd D: The determination of antithrombin-III by radioimmunoassay and its clinical application. Br J Haematol 41(4):563–5721, 1979.

Fukutake K: Progress in the determination of blood coagulation factors. Rinsho Ketsueki 20(5):462–475, 1979.

Green LH, Seroppian E, Handin RI: Platelet activation during exercise-induced myocardial ischemia. N Engl J Med 302(4):193–197, 1980.

Holmberg L, Borge L, Ljung R, Nilsson IM: Measurement of antihaemophilic factor A antigen (VIII : CAG) with a solid phase immunoradiometric method based on homologous non-haemophilic antibodies. Scand J Haematol 23(1):17–24, 1979.

Holmberg L, Nilsson IM: VIII R : AG in platelets from patients with various forms of von Willebrand's disease. Thromb Haemostasis 42(3):1033–1038, 1979.

Kierulf P, Brosstad F, Godal HC, Lund PK, Anderson AB, Naeverlid I: Poor discriminating power of fibrinogen–sepharose toward plasma fibrinogen, fibrin des-aa and RIA-FPA quantitation. Thromb Res 16(1–2):1–10, 1979.

Peake IR, Bloom AL, Giddings JC, Ludlam CA: An immunoradiometric assay for procoagulant factor VIII antigen: results in haemophilia, von Willebrand's disease, and fetal plasma and serum. Br J Haematol 42(2):269–281, 1979.

Reisner HM, Barrow ES, Graham JB: Radioimmunoassay for coagulant factor VIII–related antigen (VIII : CAG). Thromb Res 14(1):235–239, 1979.

Rucinski B, Niewiarowski S, James P, Walz DA, Dbudzynski AZ: Antiheparin proteins secreted by human platelet. Purification, characterization and radioimmunoassay. Blood 53(1):47–62, 1979.

Ruggeri ZM, Mannucci PM, Jeffcoate SL, Fyram GIC: Immunoradiometric assay of factor VIII–related antigen with observations in 32 patients with von Willebrand's disease. Br J Haematol 33:221, 232, 1976.

Saito H: Determination of Hagman factor (HG, factor XII) and plasma prekallikrein (Fletcher factor) by radioimmunoassays. Adv Exp Med Biol 120A:165–172, 1979.

Savidge GJ, Carlebjork G: An optimized radioimmunoassay of factor VIII–related antigen (F VIII R : AG) in plasma and eluates. Thromb Res 14:363–376, 1979.

Enzyme-linked Immunoassay

Carlier Y, Bout D, Capron A: Automation of enzyme-linked immunosorbent assay (ELISA). J Immunol Methods 31:237–246, 1979.

Holmberg L, Borge L, Ljung R, Milsson IM: Measurement of antihaemophilic factor A antigen (VIII : CAG) with a solid phase immunoradiometric method based on homologous non-haemophilic antibodies. Scand J Haematol 23(1):17–24, 1979.

Nel JD, Stevens K: A new method for the simultaneous quantitation of platelet-bound immunoglobulin (IgG) and complement (C_3) employing an enzyme-linked immunosorbent assay (Elisa) procedure. Br J Haematol 44:281–290, 1980.

Voller A, Bartlett A, Bidwell DE: Enzyme immunoassays with special reference to ELISA techniques. J Clin Pathol 31:507–520, 1978.

Yorde LD, Hussey CV, Yorde DE, Sasse EA: Competitive enzyme-linked immunoassay for factor VIII antigen. Clin Chem 25:1924–1927, 1979.

5

Procedure for Immunoassays

Proteins of the Coagulation System

Fibrinogen

Radial Immunodiffusion (RID) Method

Principle. The RID plates contain a mixture of agarose and a monospecific antiserum to human fibrinogen. Known concentrations of a standard are applied to a series of wells cut in the agarose. The antigen in the sample diffuses radially into the medium and reacts with the antiserum to form a visible white precipitin ring. The diameter of the ring is proportional to the amount of antigen in the sample. A standard curve of fibrinogen concentrations vs. the square of the ring diameters is plotted and the patient's sample is read off the graph.

Materials required

1. M-Partigen Fibrinogen Radial Immunodiffusion Plates (Calbiochem Behring Corp). Store in refrigerator.
2. Protein standard plasma. Reconstitute with 0.5 ml distilled water. Good for 7 days after reconstitution, when kept at 4°C.
3. Microliter pipette.
4. Measuring device, view box, or RID viewer.
5. Distilled water and normal saline.

Specimen

1. Plasma, 100 μl, esterified with citrate, EDTA, or oxalate (see Appendix).
2. Dilute specimen 1 : 2 with saline.

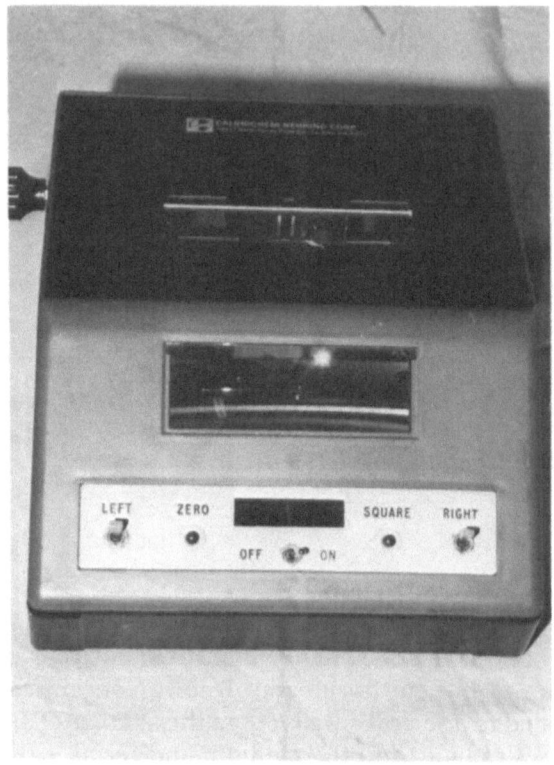

Figure 20. RID viewer for measuring diameter of the precipitate rings. The RID plate is placed on the top of the viewer.

Procedure

1. Remove RID plate from the package and let it sit open at room temperature for 5 minutes.
2. Make 1 : 2 and 1 : 4 dilutions of standard plasma in saline. Mix well and apply 5 μl of undiluted and diluted plasma to wells 1, 2, and 3.
3. Apply 5 μl of patient(s) plasma to wells 4–12.
4. Close the plate, replace in aluminum cover, and keep at room temperature in a horizontal position.
5. Keep a record of the batch number of plates and standard, date and time of sample application, concentration of standards, and the patient number for each well.
6. Measure the diameter of the precipitin ring after 48 hours, using a calibrated measuring ruler and a view box. RID viewer (Fig. 20) can also be used. Square the ring diameter readings (Fig. 21).

Figure 21. M-Partigen RID method for the measurement of fibrinogen in plasma.

7. Plot a standard curve using concentrations of the three standards vs. squared ring diameters. Draw a straight line through the points which should intercept the Y-axis at 11 ± 3.5 mm.
8. Read the concentration in patient sample from the standard curve and multiply by the dilution factor 2. Concentration values are obtained as mg/dl.

Normal range

1. 200–400 mg/dl.

Clinical significance

1. Fibrinogen is an important plasma protein (MW 340,000 daltons) and is essential for proper clot formation.
2. It consists of three polypeptide chains, termed alpha, beta, and gamma, which are linked together by disulfide bonds.
3. Thrombin digestion of fibrinogen results in two molecules of fibrinopeptide A and two molecules of fibrinopeptide B. Fibrinopeptide A levels in plasma indicate activation of the coagulation system.
4. Proteolytic action of plasmin on fibrinogen produces a small peptide, $B\beta 15\text{-}42$, from the beta chain and other fibrin(ogen) products X, Y, D, and E. These products are regarded as molecular markers for fibrinolysis.

5 Based upon immunologic characterization, many abnormal forms ("Dys-form") of fibrinogen have been identified; these Dys-forms are named after the cities in which the disorder was originally discovered.

6. Patients with a fibrinogen molecule Dys-form do not show clinical bleeding.

Fibrinopeptide A

Radioimmunoassay

Principle.   Thrombin digestion of fibrinogen results in the production of 2 fibrinopeptide A (FPA) and 2 fibrinopeptide B molecules. Fibrino-peptide A concentration in blood is quantitated to diagnose thrombosis. In a radioimmunoassay procedure, ^{125}I-labeled FPA is allowed to compete with unlabeled FPA for antibody binding sites. Free and bound fibrinopeptide A are separated. A standard curve of percent B/B_0 vs. concentration of standards is plotted on semilogarithmic paper. The quotient % B/B_0 represents part of ^{125}I-labeled FPA bound in the absence of fibrinopeptide A. FPA concentration in patient plasma is obtained from this standard curve.

Materials required

1. 12 × 75 mm polypropylene tubes.
2. Microliter pipettes (100 and 200 μl).
3. Absorbent paper.
4. Vortex mixer.
5. 1-ml pipetting syringe.
6. Timer.
7. Biogamma counting vials.
8. Biogamma counter and calculator/printer.
9. Linear graph paper.

Reagents

1. RIA-Quant FPA Test Kit (Mallinckrodt, Inc.). The FPA Test Kit contains the following reagents (the procedure is also provided with the kit).
   a. ^{125}I-Fibrinopeptide A reaction solution—yellow color.
   b. Fibrinopeptide A standards (0, 1.0, 2.0, 5.0, 10.0, 20.0, and 40.0 ng/ml).
   c. Rabbit fibrinopeptide A antiserum—blue color.
   d. Polyethylene glycol (PEG)–second antibody solution.
   e. Bentonite slurry.

f. Nonspecific binding (NSB) buffer solution—blue color.
g. Anticoagulant—reconstitute with 22 ml distilled water and store at 4°C.
h. Fibrinopeptide A. Quality controls A and B.

Specimen collection

1. Add 0.5 ml of anticoagulant to evacuated 7-ml red tube using a syringe. Do not inject air.
2. Collect venous blood without trauma or stasis. Add 5 ml of blood to the tube containing anticoagulant.
3. Mix gently and centrifuge the tube at 3000 rpm for 20 minutes.
4. Separate plasma from the cells and freeze at −20°C.

Procedure

1. Treat the plasma with bentonite as follows: Add 0.5 ml of plasma to a tube containing 1.0 ml of bentonite slurry. Mix on a vortex and centrifuge at 3000 rpm for 10 minutes. Collect supernatant with a plastic pipette and discard the sediment.
2. Arrange 3 sets of 12 × 75 mm polypropylene tubes and label as standard, control, and patient. Run each test in duplicate.
3. Add 200 μl each of standard (0–40 ng/ml), to a series of tubes. Set up separate tubes for control and patient sample and add 200 μl of specimen.
4. Add 100 μl of ^{125}I-FPA solution (yellow) to all tubes.
5. Add 100 μl of FPA antiserum (green) to all tubes.
6. Vortex each tube.
7. Cap all tubes and incubate 40 minutes to 2 hours at room temperature. (We use 1-hour incubation in our labs.)
8. Mix PEG–second antibody solution and add 1.0 ml to all tubes. Vortex.
9. Centrifuge at 3000 rpm for 20 minutes.
10. Decant the tubes and absorb the liquid over paper towels.
11. Count all tubes on gamma counter for 1 minute (Fig. 22).

Results

1. To calculate % B/B_0 use the following equation:

$$\frac{\text{CPM std, patient or control} - \text{CPM of NSB} \times 100}{\text{CPM for 0.0 } \mu\text{g/ml std} - \text{CPM of NSB}}$$

CPM = average count per minute; NSB = non-specific binding buffer.

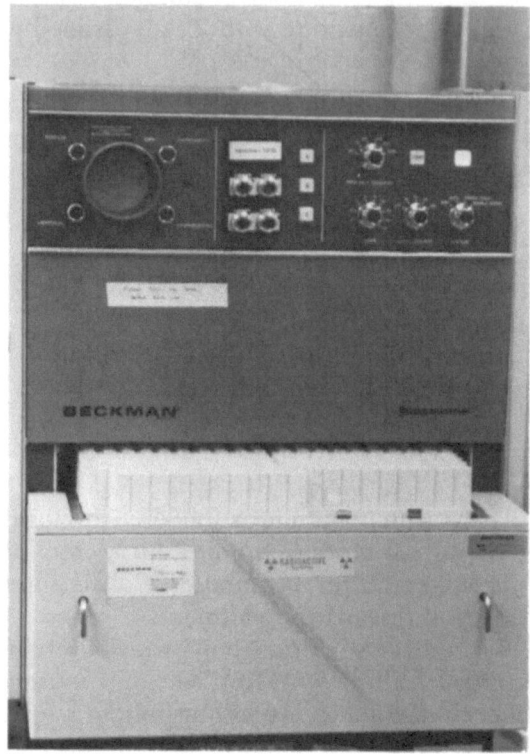

Figure 22. Counting of radioactivity in a Biogamma Counter. Courtesy of Beckman Instruments.

2. Construct a standard curve by plotting % B/B_0 vs. concentration of standard on semilogarithmic graph paper.
3. Read patient FPA concentration from the standard curve.

Normal range

1. Mean = 1.94 ± 1.2 ng/ml.
2. Range = 0.70 − 3.1 ng/ml.

Clinical significance

1. Fibrinopeptide A is a very specific marker of thrombin action on fibrinogen and its presence in plasma indicates a low grade catabo·lism of fibrinogen.
2. Its concentration in normal individuals is reported to range from 0.70 to 3.0 ng/ml.

3. Elevated levels of fibrinopeptide A are found in patients with thrombosis and other inflammatory disorders.
4. Fibrinopeptide A assays may be utilized in monitoring heparin therapy, in venous thromboembolic disease, since heparin infusions have been shown to abruptly decrease fibrinopeptide A levels.

Factor VIII : RAg

Rocket Immunoelectrophoresis

Principle.   A mixture of agarose and antiserum to Factor VIII : RAg is poured onto a glass plate and allowed to gel. A cylindrical well is cut into the agarose, and antigen sample is incorporated into the well and electrophoresed. The height of the precipitin rocket is proportional to the amount of antigen present in the sample.

Materials required

1. Power supply (Brinkman).
2. Electrophoresis chamber (Bio-Rad).
3. Glass plates, 8 × 10 cm (Scientific Products).
4. Whatman filter paper No. 1.
5. Well cutters (Gelman).
6. Pasteur pipettes.
7. Microliter pipette.

Reagents

1. Antiserum to human Factor VIII (Calbiochem Behring Corp.).
2. Seakem LE agarose (Marine Colloids).
3. Normal saline.
4. 0.05 M barbital buffer, pH 8.6.
   20.52 g sodium barbital (sodium diethylbarbiturate).
   3.68 g barbital (diethylbarbituric acid).
   2 liter distilled water.
5. Coomassie Brilliant Blue Stain (Kodak).
   Coomassie Brilliant Blue R-250, 0.50 g.
   Absolute methanol, 45.0 ml.
   Glacial acetic acid, 10.0 ml.
   Distilled water, 45.0 ml.
6. Destaining solution.
   Absolute methanol, 50.0 ml.
   Glacial acetic acid, 10.0 ml.
   Distilled water, 50.0 ml.
7. Tracking dye, bromophenol blue.

Specimen

1. Both frozen and freshly collected plasma samples can be used.
2. Dilute patient samples 1:2 and 1:4 using diluted (1:3) barbital buffer.
3. Make serial dilutions of normal human pooled plasma (NHP) or any commercial reference plasma using diluted (1:3) barbital buffer.
   Undiluted = 100 antigen units/ml
   1:2   = 50 antigen units/ml
   1:4   = 25 antigen units/ml
   1:8   = 12.5 antigen units/ml
   1:16 = 6.25 antigen units/ml
   1:32 = 3.125 antigen units/ml

Procedure

A. Precoating of glass plates.
   1. Wash the glass plates in distilled water and air dry.
   2. Precoat slides in 1% Seakem LE agarose in distilled water.
   3. Let the plates air dry.
   4. When dry, mark the top of the slides with a marker pen.
B. Preparation of agarose plates.
   1. Dissolve 0.85 g Seakem LE agarose in 100 ml of 1:3 barbital buffer, pH 8.6, and let boil.
   2. Cool the agarose to 56°C and add 100 μl of undiluted antiserum to 30 ml of melted agarose.
   3. Using a prewarmed glass pipette, transfer 15 ml of the above agarose–antiserum mixture onto a precoated glass plate.
   4. Let the agarose set at room temperature. Keep the plates covered in the refrigerator for at least 30 minutes after preparing.
   5. Using Gelman well cutters, punch a series of 10 wells in the gel 2 cm from the top of plates and 1 cm apart.
   6. Remove cut agarose by suction through a Pasteur pipette without breaking the edge of the wells.
   7. Patient samples are used as undiluted, 1:2, and 1:4.
   8. Apply 5 μl of standard dilutions into wells 1–6 and patient samples into the remaining wells. The last well contains the tracking dye, bromophenol blue.
   9. Fill each electrophoresis chamber with 1 liter of buffer.
   10. Run cool water continuously through the cooling stage at least 30 minutes prior to electrophoresis.
   11. Place plates in the chamber so that the wells are toward the cathode (−).
   12. Place dry filter paper wicks at each end of the plates and moisten with buffer.

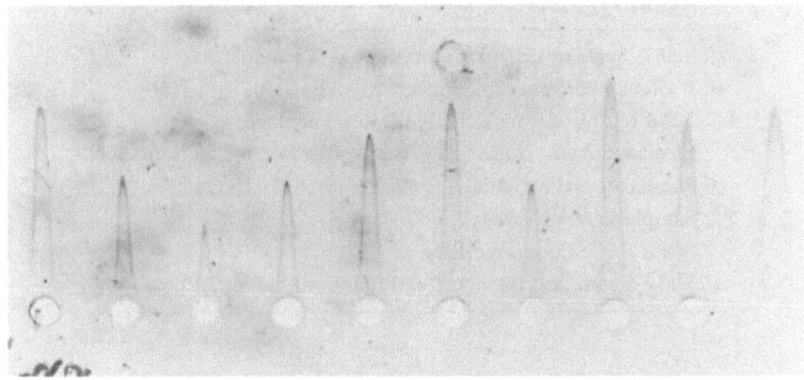

Figure 23. Measurement of Factor VIII:RAg by a rocket immunoelectrophoresis technique. The rocket height is measured from the top of the well to the rocket tip, and is plotted against the antigen concentration.

13. Press the wicks to ensure complete contact with agarose.
14. Run a current of 3 mA/plate for 30 minutes to concentrate the samples, and then switch to 5 mA/plate and electrophorese for 20 hours.
15. At the end of electrophoresis wash plates in saline for 12 hours.
16. Wash plates twice in distilled water for 1 hour each.
17. Dry the plates in an oven set at 60°C; keep a moist filter paper on the top of the plates to ensure even drying.
18. Stain the dried plates in Coomassie Blue for 6–10 minutes.
19. Destain the plates in destaining solution.
20. Wash the plates in tap water to remove excess acetic acid.
21. Measure the distance from the top of the well to the top of the peak (Fig. 23).
22. Plot a standard curve of normal pooled plasma dilutions on linear graph paper using the X-axis for antigen units and the Y-axis for rocket height in centimeters.
23. Determine the antigen concentration in patient samples from the standard curve.

Normal range

1. 50–135% of normal.

Clinical significance

1. Factor VIII is a complex of at least two high molecular weight proteins; it plays a crucial role in hemostasis.

Table 11. Assay Protocol for Factor VIII : RAg

0.85% Seakem LE agarose in diluted (1 : 3) buffer
Buffer: 0.05 M barbital buffer, pH 8.6
50 μl of anti–factor VIII : RAg/15 ml agarose
Electrophoresis: 5 mA/plate, 8.3 × 10.2 cm, for 20 hours
Standard: NHP dilutions (1 : 2, 1 : 4, 1 : 8, 1 : 16, 1 : 32)
Sample size: 5 μl/well
Stain: 0.5% Coomassie Blue
Make a standard curve of NHP dilutions vs. rocket height

---

2. Hemophilia A patients lack a coagulant protein associated with Factor VIII (Factor VIII : C) and have a normal level of the antigenic protein called Factor VIII–related antigen (VIII : RAg).
3. In von Willebrand's disease, both coagulant and antigenic protein concentrations are reduced.
4. Quantitation of Factor VIII : RAg, therefore, provides important information in the diagnosis of hemophilia A.
5. A general protocol for the rocket assay is given in Table 11.

Crossed Immunoelectrophoresis (CIEP)

Principle.   Antigen in the test plasma is first separated from other proteins by electrophoresis in an agarose gel. The separated proteins are further electrophoresed at an angle of 90° to the direction of first electrophoresis, in agarose-containing antigen-specific antiserum. This method is used to detect microheterogeneity in proteins, which is indicated by an abnormal precipitin arc or by abnormal electrophoretic migration.

Materials required

1. Power supply (Brinkman).
2. Electrophoresis chamber with cooling system (Bio-Rad).
3. Glass plates, 8 × 10 cm (Scientific Products).
4. Whatman filter paper No. 1.
5. Pasteur pipettes.
6. Surgical blade (Bard-Parker No. 11).

Reagents

1. Seakem LE agarose (Marine Colloids).
2. Antiserum to human Factor VIII (Calbiochem-Behring).

3. Coomassie brilliant blue R-250 (Kodak).
4. Bromophenol blue—marker dye (Calbiochem-Behring).
5. Absolute methanol.
6. Glacial acetic acid.
7. 0.05 M barbital buffer, pH 8.6.
   20.52 g sodium barbital (sodium diethylbarbiturate).
   3.68 g barbital, dissolved in 2 liter distilled water, pH adjusted to 8.6.

Procedure

A. Precoating of glass plates.
   1. Wash the plates in distilled water and air dry.
   2. Precoat slides in 1% agarose made in distilled water.
   3. Let the plates air dry.
   4. When dry, mark the top of the slides with a marking pen.
B. Preparation of agarose plates.
   1. Dissolve 1 g of agarose in 100 ml of diluted (1 : 3) barbital buffer, pH 8.6, and let boil.
   2. Cool the agarose to 56°C.
   3. Using a prewarmed glass pipette, transfer 15 ml of agarose onto a precoated glass plate.
   4. Let the agarose set at room temperature. Keep the plates covered in the refrigerator for at least 30 minutes after preparing.
   5. Using a Pasteur pipette, cut one well at the corner of the slide 2 cm apart from both edges.
   6. Remove cut agarose by suction without breaking the edge of the well.
   7. Apply 50 μl of the test plasma into the well.
   8. Fill the electrophoresis chambers with equal volumes of buffer. Bio-Rad Model 1512 requires 2 liters of buffer.
   9. Run cool water continuously through the cooling stage at least 30 minutes prior to electrophoresis.
   10. Place the plate in the chamber so that the well is toward the cathode (−).
   11. Place 2–4 layers of filter paper at each end of the plate; completely moisten the wicks with buffer before electrophoresis.
   12. Press wicks to ensure complete contact with agarose.
   13. Pipette 5 μl of the marker dye onto the agarose near the well.
   14. Electrophorese in the first direction at 5 mA/plate until the marker dye has reached the end of the plate.
C. Second dimension.
   1. Stop electrophoresis and move the plate at a 90° angle.
   2. Switch the buffers in the chamber.
   3. With a surgical blade, make a longitudinal cut near the well and remove the agarose from the right side of the plate (see Fig. 11).

4. Make a 0.9% agarose solution in diluted buffer, cool to 56°C, and incorporate 0.1 ml of undiluted anti–Factor VIII R : Ag with 6 ml of agarose.
5. Pour 5 ml of agarose–antibody mixture on remaining surface of the slide. Keep the plate in the refrigerator for 15 minutes.
6. Place filter paper wicks on each end of the plate. Electrophorese 16 hours with 5 mA/plate.

D. Washing and staining.
1. At the end of the electrophoresis wash the plate in saline for 2–3 hours.
2. Rinse the plate in distilled water for 1 hour.
3. Dry the plates in an oven set at 60°C; keep a moist filter paper on the top of the plates to ensure even drying. The plates can be dried overnight at room temperature.
4. Prepare a 0.50% Coòmassie Blue stain (0.50 g stain, 45 ml methanol, 10 ml acetic acid, make up to 100 ml with distilled water).
5. Stain the dried plates in Coomassie Blue for 6–10 minutes, or longer.
6. Destain the plates, using a 5 : 5 : 1 mixture of methanol : distilled water : glacial acetic acid as destaining solution, until the precipitin arc becomes distinct.
7. Wash the plates in tap water to remove excess acetic acid.

E. Evaluation.
1. Normal plasma produces a single peak and migrates 1–2 cm from the center of the well.
2. Any abnormality in Factor VIII : RAg is indicated by more than one peak.

Fluoroimmunoassay (FIA)

Principle.   Plasma containing Factor VIII : RAg is reacted with fluorescein-labeled anti–Factor VIII : RAg. Excess of unbound labeled antibody is then allowed to react with purified Factor VIII : RAg adsorbed on a cellulose disc mounted on an inert plastic handle, i.e., a Stiq sampler. The fluorescence on the Stiq sampler is measured in a fluorometer and is inversely proportional to the concentration of Factor VIII : RAg in the plasma.

Materials required

1. FIAX System (IDT).
   a. Fluorometer.
   b. Dilutor.

Figure 24. Components of the FIAX system. Courtesy of IDT.

   c. Microprocessor.
   d. Horizontal shaker.
   e. Stiq samplers.
   f. Test tube rack.
2. Microliter pipettes (25, 50, and 500 μl).
3. 12 × 75 mm borosilicate culture tubes.
4. Parafilm.
5. Masking tape.

Reagents

1. Antihemophilic factor (human), lyophilized; (Alpha Therapeutic).
2. FITC-labeled antiserum to Factor VIII : RAg (Atlantic Antibodies).
3. Factor VIII : RAg standard (Diagnostica Stago).
4. Phosphate buffered saline, 0.05 M, pH 7.4, containing 4% bovine serum albumin: the reaction buffer.
5. Phosphate buffered saline, pH 7.7, containing 2% bovine serum albumin: the measurement buffer.

Specimen

1. Fresh or frozen citrated plasma.

Procedure

A. Preparation of Stiq samplers.
    1. Reconstitute lyophilized antihemophilic factor (human) in reaction buffer and adjust the concentration to 20 units of Factor VIII : RAg/ml to serve as a stock solution.
    2. Make a 1 : 100 dilution of the above stock solution in the reaction buffer. This dilution will result in 0.2 units of Factor VIII : RAg/ml.
    3. Arrange 50–100 Stiq samplers in a row and place 25 μl of Factor VIII : RAg solution (0.2 unit/ml) onto the cellulose nitrate disc surface. Allow the Stiq samplers to dry overnight at room temperature and store in plastic bags at 4°C. The precoated samplers are good for 2–3 months.
B. FITC-labeled anti–Factor VIII : RAg.
    1. Make a 1 : 300 dilution of the antiserum by using reaction buffer.
    2. Each test requires 500 μl of the antiserum. Make as much antiserum as needed.
    3. The dilution of antiserum depends on the titer of the antiserum used.
C. Calibrators.
    1. Reconstitute Factor VIII : RAg standard lyophilized powder in 0.5 ml distilled water.
    2. Make 50% and 25% dilutions by using the reaction buffer.
    3. A 4-point calibration curve consists of the following calibrator concentrations.

    Cal I   =   25%
    Cal II =   50%
    Cal III = 100%
    Cal IV = 200%

    4. A 100% preparation is assumed to contain 1 unit of Factor VIII : RAg/ml.
D. Assay protocol.
    1. Arrange two rows of 12 × 75 mm test tubes in a rack.
    2. Dispense 500 μl of diluted (1 : 300) antiserum into each tube in the first row.
    3. Dispense 500 μl of measurement buffer into each tube in the second row.
    4. Add 50 μl of Cal I into tube #1 in the first row and mix gently.
    5. Add 50 μl of Cal II and III into tubes #2 and #3, respectively. Add 100 μl of Cal IV into tube #4 of the first row and gently mix the solution.
    6. Place one precoated Stiq sampler into each tube in the first row, and face the cellulose nitrate surface downward.

7. Place the rack on the horizontal shaker and shake the tubes for 20 minutes.
8. Stop the shaker after 20 minutes, transfer the Stiq samplers into corresponding tubes in the back row, and shake for another 20 minutes.
9. Stop the shaker.
10. Turn "ON" the fluorometer and microprocessor.
11. Follow these instructions with the microprocessor:
ENTER DATE
DATE: 10/21/81 (example)
ENTER OPER ID
OPID: 1/300 (for antibody dilution)
ENTER TEST NUMBER
TEST: C4 (example)
ENTER REPLICTS 1–3
REPL: 1
ENTER CAL 1 VALUE
CAL 1 25.0 MG/DL
ENTER CAL 2 VALUE
CAL 2 50.0 MG/DL
ENTER CAL 3 VALUE
CAL 3 100.0 MG/DL
ENTER CAL 4 VALUE
CAL 4 200.0 MG/DL
INSERT CALIBRTR 1 (at this point, insert Stiq sampler from tube #1)
INSERT CALIBRTR 2 (Stiq sampler from tube #2)
INSERT CALIBRTR 3 (Stiq sampler from tube #3)
INSERT CALIBRTR 4 (Stiq sampler from tube #4)

CAL NUM	FSU	CONC (MG/DL)
1	110	25
2	57	50
3	92	100
4	89	200

CAL NUM	DEV	CON (MG/DL)
1	−51%	12
2	166%	133
3	24%	124
4	−37%	125

ACCEPT: PUSH ENTER
OMIT: ENTER CAL NUM (enter 2 at this point)

****

DATE: 10/21/81
TEST: C4
REPL: 1

CAL NUM	FSU	CONC (MG/DL)
1	110	25
2	57	50
3	92	100
4	89	200

CAL NUM	DEV	CONC (MG/DL)
1	−1%	25
2	OMITTED	
3	5%	105
4	−3%	194

SAMP NUM	FSU	CON (MG/DL)
001	Insert Stiq sampler for patients	
002	#1, 2, 3, etc.	
003		
004		
005		

Prothrombin

Rocket Immunoelectrophoresis

Principle.   Antiserum to human prothrombin is incorporated into an agarose medium and cylindrical wells are cut in the gel matrix. Plasma containing prothrombin is applied to the well and the gel is subjected to electrophoresis. Antigen–antibody reaction results in the formation of rocket-shaped structures. The height of the rockets is directly proportional to the amount of prothrombin present in the plasma.

Materials required

1. Power supply (Brinkman).
2. Electrophoresis chamber (Bio-Rad).
3. Glass plates, 8 × 10 cm (Scientific Products).
4. Well cutters (Gelman).
5. Pasteur pipettes.
6. Microliter pipettes.
7. Whatman filter paper No. 1.

Reagents

1. Antiserum to human prothrombin (Nordic).
2. Seakem LE agarose (Marine Colloids).
3. Normal saline.
4. 0.05 M barbital buffer, pH 8.6.
   20.52 g sodium barbital (sodium diethylbarbiturate).
   3.68 g barbital (diethylbarbituric acid).
   2 liter distilled water.
5. Coomassie Brilliant Blue Stain (Kodak).
   Coomassie Brilliant Blue R-250, 0.50 g.
   Absolute methanol, 45.0 ml.
   Glacial acetic acid, 10.0 ml.
   Distilled water, 45.0 ml.
6. Destaining solution.
   Absolute methanol, 50.0 ml.
   Glacial acetic acid, 10.0 ml.
   Distilled water, 50.0 ml.

Specimen

1. Both frozen and freshly collected plasma samples can be used.
2. Dilute patient samples 1 : 2 using saline.
3. Make serial dilutions of normal human pooled plasma (NHP) or any other commercial reference plasma as follows:
   Undiluted = 100%
   1 : 2 =   50%
   1 : 4 =   25%
   1 : 8 =   12.25%

Procedure

1. Dissolve 0.85 g Seakem agarose in 100 ml of 1 : 3 diluted buffer, and let boil.

2. Cool the agarose to 56°C and add 150–200 μl of undiluted antiprothrombin to 15 ml of agarose.
3. Transfer 10–12 ml of this mixture onto the precoated glass surface. (For precoating, see procedure for Factor VIII : RAg.)
4. Keep the plates in the refrigerator for 30 minutes.
5. Using gel cutters, punch a series of 10 wells in the gel 2 cm from the top of the plate and 1 cm apart.
6. Carefully remove agarose by suction through a Pasteur pipette.
7. Apply 5 μl of serial NHP dilutions into wells 1–4, respectively, and patient samples into the remaining wells. The last well contains the tracking dye, bromophenol blue.
8. Fill the chamber with 1 liter of buffer in each compartment.
9. Place the plates on the stage of the chamber with the wells toward the cathode. Place paper wicks on the agarose at both cathodic and anodic sides.
10. Run a constant current of 3–5 mA/plate for 16–18 hours.
11. Wash plates in saline for 1–2 hours and twice in distilled water for 30 minutes each.
12. Dry the plates in an oven set at 60°C; keep a wet paper towel on the top of the plates.
13. Stain the dried plates in Coomassie Blue for 6–10 minutes.
14. Destain the plates in destaining solution.
15. Wash the plates in tap water to remove excess acetic acid.
16. Measure the distance from the top of the well to the top of the rocket peak.
17. Plot a standard curve of plasma dilutions vs. rocket height; use the curve to determine prothrombin concentration in patient plasma.

Results

1. Range = 70–130% of normal.

Clinical significance

1. Prothrombin is a vitamin K–dependent protein with a molecular weight of 72,000 daltons. Its concentration in plasma varies from 5 to 10 mg/dl.
2. Prothrombin is acted upon by Factor Xa-V complex which results in the production of thrombin. Fragments of prothrombin are also produced during the conversion of prothrombin to thrombin.
3. Congenital prothrombin deficiency is reported in some 26 families, so far.

Factor XII (Hageman Factor)

Rocket Immunoelectrophoresis

Principle.   A highly purified agarose solution containing Factor XII antiserum is poured onto a glass plate and allowed to gel. Patient plasma containing Factor XII is incorporated into a series of wells and electrophoresis is performed. Rockets due to antigen–antibody reaction are measured. The height of the rocket is directly proportional to Factor XII concentration in the plasma.

Materials required

1. Power supply (Brinkman).
2. Electrophoresis chamber (Bio-Rad).
3. Glass plates, 8 × 10 cm (Scientific Products).
4. Well cutters (Gelman).
5. Pasteur pipettes.
6. Microliter pipettes.

Reagents

1. Antiserum to Factor XII (Nordic).
2. Seakem LE agarose (Marine Colloids).
3. Normal saline.
4. 0.05 M barbital buffer, pH 8.6.
   20.52 g sodium barbital (sodium diethylbarbiturate).
   3.68 g barbital (diethylbarbituric acid).
   2 liter distilled water.
5. Coomassie Brilliant Blue Stain (Kodak).
   Coomassie Brilliant Blue R-250, 0.50 g.
   Absolute methanol, 45.0 ml.
   Glacial acetic acid, 10.0 ml.
   Distilled water, 45.0 ml.
6. Destaining solution.
   Absolute methanol, 50.0 ml.
   Glacial acetic acid, 10.0 ml.
   Distilled water, 50.0 ml.

Specimen

1. Both frozen and freshly collected plasma samples can be used.
2. Dilute patient samples 1 : 2 using saline.

3. Make serial dilutions of normal human pooled plasma (NHP) or any commercial reference plasma as follows:
   1 : 32
   1 : 64
   1 : 128

Procedure

A. Precoating of glass plates.
   1. Wash the glass plates in distilled water and air dry.
   2. Precoat plates in 1% Seakem LE agarose made in distilled water.
   3. Let the plates air dry.
   4. When dry, mark the top with a marker pen.
B. Preparation of agarose plates.
   1. Dissolve 0.85 g Seakem agarose in 100 ml of 1 : 3 diluted buffer, and let boil.
   2. Cool the agarose to 56°C and add 300 μl of anti–Factor XII in 10 ml of agarose.
   3. Transfer 10–12 ml of antiserum–agarose mixture onto the precoated glass surface.
   4. Let the agarose settle at room temperature. Keep the plates covered in a refrigerator for at least 30 minutes after preparation.
   5. Using Gelman gel cutters, punch a series of 10 wells in the gel 2 cm from the top of the plate and 1 cm apart.
   6. Carefully remove agarose by suction through a Pasteur pipette.
   7. Apply 5 ml of the NHP dilutions into wells 1–3, respectively, and patient samples into the remaining wells. The last well contains the tracking dye, bromophenol blue.
   8. Fill the chamber with 1 liter of buffer in each compartment.
   9. Place the plate on the stage of the chamber, so that the wells are toward the cathode. Make sure that the plates are in good contact with the wicks.
   10. Run a current of 3–5 mA/plate for 16–20 hours.
   11. At the end of electrophoresis, wash plates in saline for 12 hours and then twice in distilled water for 1 hour each.
   12. Dry the plates in an oven set at 60°C; keep a moist paper towel on top of the plates to ensure even drying.
   13. Stain the dried plates in Coomassie Blue for 6–10 minutes.
   14. Destain the plates in destaining solution.
   15. Wash the plates in tap water to remove excess acetic acid.
   16. Measure the distance from the top of the well to the top of the rocket peak (Fig. 25).
   17. Plot a standard curve of plasma dilutions vs. rocket height and determine Factor XII concentration in patient plasma from this curve.

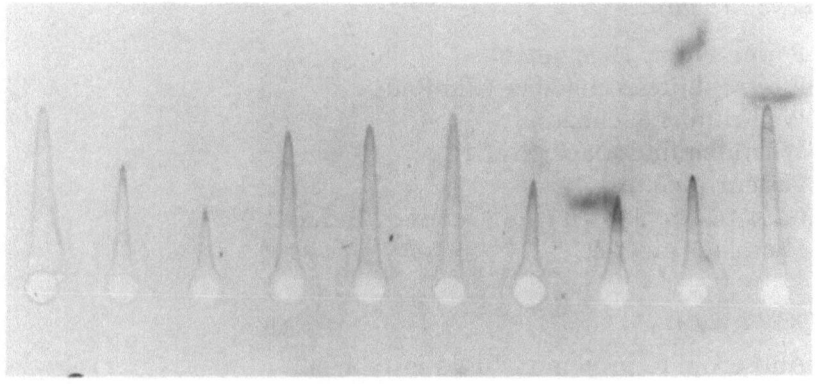

Figure 25. A typical rocket formation in an electroimmunodiffusion method for Factor XII (Hageman factor).

Clinical significance

1. Hageman factor is a single chain protein with a molecular weight of 90,000 daltons. Its concentration in plasma is reported to be 27–45 μg/liter.
2. It can be activated by negatively charged surfaces such as glass, kaolin, celite, dextran sulfate, ellagic acid, collagen, and other vascular wall components.
3. The fibrinolytic enzyme plasmin is also known to activate Hageman factor into its fragments.
4. A deficiency of Factor XII is not usually associated with clinical bleeding.

Protein C*

Rocket Immunoelectrophoresis

Principle.   A monospecific antiserum to protein C raised in rabbits is mixed with an agarose medium and allowed to gel on a glass plate. Purified protein C antigen concentrations are applied to a series of wells cut in the gel matrix and electrophoresis is performed. Rockets formed as a result of antigen–antibody reaction are measured. Height of the rocket is proportional to protein C concentration.

---

* From Fareed J, Parvez Z, Messmore HL, et al: Clinical and Experimental Studies on the Activation of Fibrinolytic Systems by Bypass Complexes: Role of Protein C in the Mediation of Fibrinolysis. Presented at the IXth International Congress on Thrombosis and Haemostasis, Stockholm, July 3–8, 1983.

Materials required

1. Power supply (Brinkman).
2. Electrophoresis chamber (Bio-Rad).
3. Well cutters (Gelman).
4. Whatman filter paper No. 1.
5. Pasteur pipettes.
6. Glass plates, 8 × 10 cm (Scientific Products).
7. Microliter pipettes.

Reagents

1. Antiserum to protein C raised in rabbits.
2. Seakem LE agarose (Marine Colloids).
3. Purified protein C (Stago).
4. Normal saline.
5. 0.05 M barbital buffer, pH 8.6.
   20.52 g sodium barbital (sodium diethylbarbiturate).
   3.68 g barbital (diethylbarbituric acid).
   2 liter distilled water.
6. Coomassie Brilliant Blue Stain (Kodak).
   Coomassie Brilliant Blue R-250, 0.50 g.
   Absolute methanol, 45.0 ml.
   Glacial acetic acid, 10.0 ml.
   Distilled water, 45.0 ml.
7. Destaining solution.
   Absolute methanol, 50.0 ml.
   Glacial acetic acid, 10.0 ml.
   Distilled water, 50.0 ml.

Specimen

1. Both frozen and freshly collected plasma samples can be used.
2. Make serial dilutions of protein C standard to obtain concentrations of 10, 5, 2.5, 1.25, and 0.625 μg/ml.

Procedure

A. Precoating of glass plates.
   1. Wash the glass plates in distilled water and air dry.
   2. Precoat plates in 1% Seakem LE agarose made in distilled water.
   3. Let the plates air dry.
   4. When dry, mark the top with a marker pen.
B. Preparation of agarose plates.
   1. Dissolve 0.85 g Seakem agarose in 100 ml of 1 : 3 diluted buffer, and let boil.

2. Cool the agarose to 56°C and add 100 μl of anti–protein C in 10 ml of agarose.

3. Transfer antiserum–agarose mixture onto the precoated glass surface.

4. Let the agarose settle at room temperature. Keep the plates covered in a refrigerator for at least 30 minutes after preparation.

5. Using Gelman gel cutters, punch a series of 10 wells in the gel, 2 cm from the top of the plate and 1 cm apart.

6. Carefully remove agarose by suction through a Pasteur pipette.

7. Apply 5 μl of the protein C dilutions into wells 1–5, respectively, and patient samples into the remaining wells. The last well contains the tracking dye, bromophenol blue.

8. Fill the chamber with 1 liter of buffer in each compartment.

9. Place the plate on the stage of the chamber, so that the wells are toward the cathode. Make sure that the plates are in good contact with the wicks.

10. Run a current of 5 mA/plate for 20 minutes to concentrate the samples, and then increase it to 16 mA/plate for 5 hours.

11. At the end of electrophoresis, wash plates in saline for 12 hours, and then twice in distilled water for 1 hour each.

12. Dry the plates in an oven set at 60°C; keep a moist paper towel on top of the plates to ensure even drying.

13. Stain the dried plates in Coomassie Blue for 6–10 minutes.

14. Destain the plates in destaining solution.

15. Wash the plates in tap water to remove excess acetic acid.

16. Measure the distance from the top of the well to the top of the rocket peak (Fig. 26).

17. Plot a standard curve of plasma dilutions vs. rocket height and determine protein C concentration in patient plasma from this curve.

Clinical significance

1. Protein C is a vitamin K–dependent protein with a molecular weight of 56,000 daltons.

2. Its concentration in plasma ranges from 10 to 12 μg/ml.

3. The exact physiologic role of protein C is not yet known. However, the activated form has been shown to inhibit Factor V and Factor VIII.

4. Recent reports indicate that activated protein C in bypass complexes may be responsible for fibrinolysis. This action may be mediated via the release of endogenous plasminogen activators.

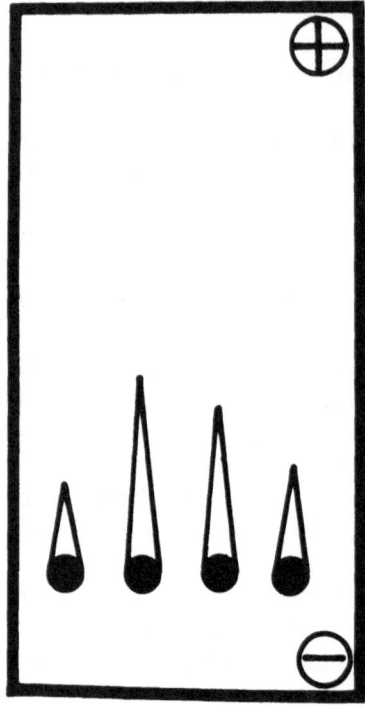

Figure 26. Measurement of protein C by a rocket immunoelectrophoresis method.

Fibronectin

Turbidimetric Immunoassay

Principle.   Known amounts of fibronectin are reacted with a monospecific antibody and the turbidity of the antigen–antibody complexes is measured in a spectrophotometer at 340 nm. A standard curve of concentration vs. OD_{340} is plotted and the fibronectin concentration in unknown plasma samples is determined from this curve.

Materials required

1. 0.05 M, pH 7.4 phosphate buffer containing 2.5% polyethylene glycol-6000 (PEG).
2. Fibronectin standard, 100–500 μg/ml (Cappel).
3. Antifibronectin IgG fraction (Cappel).
4. 12 × 75 mm glass tubes.
5. Spectrophotometer and microcuvettes with 1-cm light path.

Specimen

1. Plasma collected in 3.8% citrate or in EDTA.
2. Plasma should be frozen, if not used immediately.
3. Perform assay within 4 hours of collection.

*Procedure**

1. Arrange two sets of tubes in a rack and mark the first row as *blank* and the second row as *test*.
2. The first five tubes are marked as 100, 150, 200, 300, and 500 μg/ml standard.
3. Mark tubes 6, 7, 8, 9 . . . n as patient 1, patient 2 . . . patient n.
4. Dispense 1 ml of buffer to all the tubes in the first row (blanks).
5. Add 1 ml of diluted (1 : 10) antifibronectin antibody to all the tubes in the second row (tests).
6. Pipette 10 μl of 100, 150, 200, 300, and 500 μg/ml standard into tubes 1, 2, 3, 4, and 5 (blanks and tests).
7. Add 10 μl of patient(s) plasma(s) to tubes 6, 7, 8, 9 . . . n (blanks and tests).
8. Mix these solutions and incubate at room temperature for 10 minutes. Time should be allowed between mixing antigen with the antibody—a 2-minute interval between the tubes is usually enough.
9. Set the spectrophotometer to read at 340 nm and zero the machine (Fig. 27).
10. Transfer the contents of tube 1-blank to the microcuvette and bring the optical density reading to zero.
11. Repeat with tube 1-test and *record* optical density. Do not bring it to zero.
12. *Record* OD_{340} for tubes 2, 3, 4, 5, 6, 7 . . . n. Test, zeroing the machine with their respective blanks.
13. Using linear graph paper, plot concentration of the standard against optical densities (Fig. 28).
14. Determine fibronectin concentration in patient plasma by using this standard curve.

Normal range

1. 269–320 μg/ml.

---

* This procedure is modified from the one given by Cappel Laboratories.

Figure 27. Hitachi Digital Spectrophotometer Model 102 utilized for measuring turbidity in fibronectin immunoassay.

Clinical significance

1. Fibronectin is a high molecular weight glycoprotein which is found in blood and other body fluids.
2. It probably functions as an adhesive and as an opsonic protein.
3. It acts as a substrate for thrombin, plasmin, and Factor VIII.
4. It enhances activation of plasminogen by urokinase.

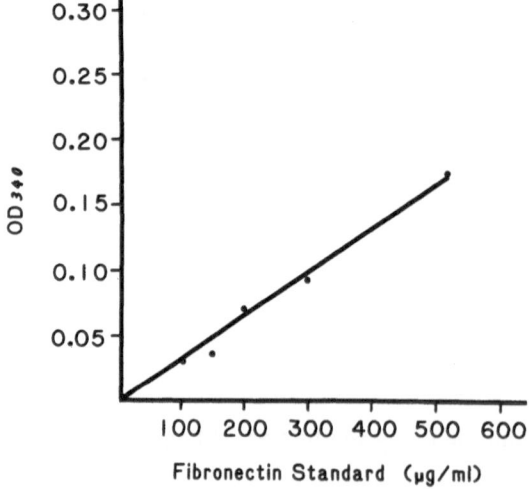

Figure 28. A typical standard curve of a turbidimetric immunoassay for fibronectin.

Table 12. Protocol for a Turbidimetric Immunoassay of Fibronectin in Plasma

Reagents	Row 1: Tube # (Blanks)								Row 2: Tube # (Tests)							
	1	2	3	4	5	6	7	8 . . .	1	2	3	4	5	6	7	8 . . .
Buffer (1 ml)	+	+	+	+	+	+	+	+								
Antibody (1 ml)									+	+	+	+	+	+	+	+
Antigen (10 µl) Standard, patient	+	+	+	+	+	+	+	+	+	+	+	+	+	+	+	+

1. Incubate each tube for 10 minutes at room temperature and read OD340 of the test sample, after zeroing the spectrophotometer reading with the blanks.
2. Construct a standard curve of reference concentration vs. OD340 and determine fibronectin concentration in unknowns from this curve.

5. It is also found in alpha-granules of platelets and may mediate adhesion of platelets to collagen.
6. It is synthesized by endothelial cells and may contribute to the thrombogenicity of subendothelial connective tissue.
7. Fibronectin concentration is decreased in disseminated intravascular coagulation (DIC).
8. A general protocol for fibronectin determination is shown in Table 12.

Suggested readings

Mosher DF: Fibronectin: relevance to hemostasis and thrombosis. In Colman RS, Hirsh J, Marder VJ, Salzman EW (eds): Hemostasis and Thrombosis, Basic Principles and Clinical Practice. Philadelphia, Lippincott, 1982, pp 174–184.

Parvez Z, Messmore HL, and Moncada R: A rapid immunoturbidimetric assay for the quantitation of fibronectin (FN) in plasma. Blood, November, 1983.

Proteins of the Fibrinolytic System

$B\beta_{15-42}$ Peptide

Radioimmunoassay*

Principle.   A constant amount of iodinated $B\beta_{15-42}$ peptide is incubated with ethanol-precipitated plasma and anti–$B\beta_{15-42}$. The antigen–antibody complex is then adsorbed to insolubilized anti–rabbit IgG, washed,

---

* This radioimmunoassay is for research use only.

and its radioactivity determined. If $B\beta_{15-42}$ is present in patient plasma, it will inhibit iodinated $B\beta_{15-42}$ peptide from binding to the first antibody. A standard curve of $B\beta_{15-42}$ concentration vs. percentage inhibition of binding is constructed and the concentration of the peptide in patient plasma is determined from this curve.

Materials required

1. Styrene centrifuge tubes.
2. Microliter pipettes.
3. Magnetic stirring bars.
4. Small magnetic stirrer.
5. Conical glass tube (3 ml).
6. Sephadex G-10 (Pharmacia).
7. Econo-column (Bio-Rad).
8. Pasteur pipettes.
9. Plastic tubes with caps.
10. Tube rack.
11. 25-μl Hamilton micropipette.
12. $Na^{125}I$, mCi carrier (New England Nuclear).
13. Gamma counter.

Reagents

1. IMCO RIA Kit for $B\beta_{15-42}$, which contains the following:
   a. Rabbit antiserum to human $B\beta_{15-42}$.
   b. Human $B\beta_{15-42}$ standard.
   c. Human $B\beta_{15-42}$ peptide for iodination.
2. RIA grade bovine serum albumin (BSA) (Sigma).
3. Buffer I (iodination buffer).
   0.5 M Na_2HPO_4, adjust pH to 7.5 with HCl.
4. Buffer II (column buffer).
   0.05 M Tris, 0.1 M NaCl, adjust pH to 7.5 with HCl.
5. Buffer III (tracer dilution buffer).
   Same as buffer II + 1 mg/ml BSA.
6. Chloramine-T (mg/ml) (New England Nuclear).
   Dissolve in water immediately before use.
7. Sodium metabisulfate ($Na_2S_2O_5$), 1 mg/ml.
   Dissolve in water immediately before use.
8. Sephadex G-10, 10 ml packed in Econo-column.

Radioiodination procedure. The following procedure is performed in a fume hood suitable for radioactive isotope work.

1. Add 25 μl B$\beta_{15\text{-}42}$ peptide to the conical glass tube with a magnetic stirrer.
2. Keep the following reagents ready:
   a. Chloramine-T, 10 μl.
   b. Na$_2$S$_2$O$_5$, 100 μl.
   c. Buffer III, 100 μl.
   d. Buffer III, 100 μl.
3. Dispense Na^{125}I to the B$\beta_{15\text{-}42}$ peptide.
4. Mix chloramine-T for 30 seconds.
5. Add Na$_2$S$_2$O$_5$ and mix for 30 seconds.
6. Apply the reaction mixture to the Sephadex G-10 column with a Pasteur pipette.
7. Wash reaction tube with 100 μl buffer III and apply to the column.
8. Elute column with buffer III at a flow rate of 6–8 ml/hour.
9. Collect 30 fractions containing 0.5 ml each.
10. Dilute 10 μl of each fraction with 1 ml water for counting.
11. Plot radioactivity profile of the fractions. Two distinct peaks should be obtained.
12. The first peak contains the tracer and fractions showing high activity are pooled.
13. Mix the tracer peak with an equal volume of buffer III.
14. Divide the tracer mixture into 25- or 50-μl aliquots and store at $-70°$C. The tracer is stable for at least 6 weeks.

Assay

Materials required

1. Micropipettes.
2. Plastic tubes with caps (3 ml).
3. Tube rack.
4. Rotation mixer.
5. Gamma counter.
6. Centrifuge.

Reagents

1. Anti–B$\beta_{15\text{-}42}$ serum from the kit.
2. Standard B$\beta_{15\text{-}42}$, 15 ng/mg in buffer IV.
3. Buffer IV (antiserum and tracer dilution buffer).
   0.04 M Tris, 0.11 M NaCl, 0.01 M EDTA,
   0.02% NaN$_3$, Trasylol (20 KIU/ml),
   RIA-grade BSA (1 mg/ml). Adjust pH to 7.4 with HCl.

Table 13. A Summary of Radioimmunoassay Procedure for Bβ₁₅₋₄₂ Peptide

Tube No.	Sample	Buffer (μl)	Bβ Standard (μl)	Plasma Sample (μl)	Antiserum (μl)	Tracer (μl)	DASP* (μl)
1–2	Buffer control	200	–	–	–	50	–
3–4	DASP control	200	–	–	–	50	400
5–6	Antiserum control	100	–	–	100	50	400
7–8	Bβ15-42 1:1	–	100	–	100	50	400
9–10	1:2	–	100	–	100	50	400
11–12	1:4	–	100	–	100	50	400
13–14	1:8	–	100	–	100	50	400
15–16	1:16	–	100	–	100	50	400
17–18	1:32	–	100	–	100	50	400
19–20	1:64	–	100	–	100	50	400
21–22	Patient 1 1:2	–	–	100	100	50	400
23–24	1:5	–	–	100	100	50	400
25–26	1:10	–	–	100	100	50	400
27–28	Patient 2 1:2	–	–	100	100	50	400
29–30	1:5	–	–	100	100	50	400
31–32	1:10	–	–	100	100	50	400

* The assay utilizes a second antibody to separate antigen–antibody complex from free antigen. Double antibody solid phase (DASP) is utilized for this purpose.

4. Buffer V (double antibody solid phase [DASP] dilution buffer). Same as buffer IV but BSA concentration is 20 mg/ml.
5. Buffer VI (standard and sample dilution buffer).
0.04 M Tris, 0.11 M NaCl, 0.01 M EDTA,
0.01% NaN$_3$, Trasylol (20 KIU/ml),
RIA-grade BSA, 25% ethanol. Adjust pH to 7.4 with HCl.
6. Conjugated sheep anti–rabbit IgG (DASP) (New England Nuclear).

Procedure. The actual radioimmunoassay procedure is summarized in Table 13.

1. Label tubes 1–32 and place them in a rack.
2. Add 200 μl buffer IV to tubes 1–4 and 100 μl to tubes 5–6.
3. Add 100 μl of B$\beta_{15\text{-}42}$ standard (15 ng/ml) to tubes 7 and 8.
4. Add 100 μl of serially diluted (1:2, 1:4, 1:8, 1:16, 1:32, and 1:64) B$\beta_{15\text{-}42}$ standard to tubes 9–10, 11–12, 13–14, 15–16, 17–18, and 19–20, respectively. These dilutions are made in buffer VI.
5. Make 1:2, 1:5, and 1:10 dilutions of patient defibrinated plasma using buffer IV and add 100 μl of these dilutions to tubes 21–22, 23–24, and 25–26, respectively (each patient sample is tested in three dilutions, i.e., 1:2, 1:5, and 1:10).
6. Repeat the same procedure for other patient plasmas.
7. Add 50 μl of diluted B$\beta_{15\text{-}42}$ tracer (25,000 cpm/50 μl) to *all* the tubes.
8. Add 100 μl of 1:10 diluted antiserum to all tubes *except* 1–4.
9. Mix and cap tubes and incubate overnight at 4°C.
10. Centrifuge 3 ml of DASP suspension, resuspend the pellet in 30 ml of buffer V, and add 400 μl to all tubes except 1–2.
11. Cap all tubes and place in the rotation mixer for 2 hours at room temperature.
12. After 2 hours, tubes containing DASP are centrifuged and the supernatant is discarded.
13. Wash DASP pellet with 1 ml of 0.15 M NaCl and repeat washing twice. Discard the supernatant.
14. Count each tube in a gamma counter for 1–2 minutes.

Calculations

1. Calculate average of count per minute (CPM) doubles. Subtract the value of the DASP control (average of tubes 3 and 4) from all averages except in NSB estimation.
2. $\dfrac{\text{DASP control}}{\text{buffer control}} \times 100 = \%$ NSB of tracer
3. $\dfrac{\text{antiserum control}}{\text{buffer control}} \times 100 = \%$ tracer bound by antiserum (% TB)

Table 14. Steps in Calculating $B\beta_{15\text{-}42}$ Peptide Concentration by Radioimmunoassay*

Tube No.	Sample	CPM (average of doubles)	CPM (minus DASP control)	NSB (%)	TB (%)	Binding of Antiserum Control (%)	Inhibition of Binding (%)	Logit	$B\beta_{15\text{-}42}$ (pmole)	$B\beta_{15\text{-}42}$ (pmole/ml plasma)
1–2	Buffer Control	22,246	21,721							
3–4	DASP Control	525		2.36	30.9					
5–6	Antiserum Control	7,231	6,706 (B₀)							
7–8	$B\beta_{15\text{-}42}$ 1:1	1254	729			11	89	−0.91		
9–10	1:2	1976	1451			22	78	−0.56		
11–12	1:4	2650	2125			32	68	−0.33		
13–14	1:8	4005	3480			52	48	+0.03		
15–16	1:16	5242	4717			70	30	+0.38		
17–18	1:32	6110	5585			83	17	+0.70		
19–20	1:64	6836	6311			94	6	+1.20		
21–22	Patient 1 1:10	4402	3877			58	42	+0.14	0.05	10 } 10.1
23–24	1:20	5468	4943			74	26	+0.45	0.025	10.25 } 10.1
25–26	Patient 2 1:10	3639	3114			46	54	−0.06	0.08	15.75 } 16.4
27–28	1:20	4707	4182			62	38	+0.22	0.04	17 } 16.4

* See package insert, $B\beta_{15\text{-}42}$ peptide for further details of calculation.

4. Values obtained in #1 above are called B.
   Values obtained in #1 above for antiserum control are called Bo. B is converted to logit by the following transformation:

$$\text{Log}_{10} \frac{\dfrac{B}{Bo}}{1 - \dfrac{B}{Bo}} = \text{logit}$$

When logit for each standard dilution is plotted vs. $\log_{10}$ of $B\beta_{15\text{-}42}$ concentration, a straight line is obtained.

Concentration of $B\beta_{15\text{-}42}$ in patient plasma is obtained from this curve, multiplied by 2 for ethanol precipitation, and further multiplied by the dilution factor. Concentration is then expressed as pmoles/ml. The calculation procedure is summarized in Table 14.

Normal range

1. 0.43 ± 0.15 pmole/ml.

Clinical significance

1. Fibrin degradation by plasmin results in the production of a small molecular weight fragment, the $B\beta_{15\text{-}42}$ peptide.
2. The presence of $B\beta_{15\text{-}42}$ peptide in plasma indicates fibrino(geno)-lysis.
3. Elevated levels of $B\beta_{15\text{-}42}$ peptide have been reported in thrombosis and DIC syndrome.

Suggested readings

IMCO Corporation: IMCO Radioimmunoassay for $B\beta_{15\text{-}42}$ Peptide. Package insert.

Kudryk B, Robinson D, Netre C, Hessel B, Blomback M, Blomback B: Measurement in human blood of fibrinogen/fibrin fragments containing the $B\beta_{15\text{-}42}$ sequence. Thromb Res 25:277–291, 1982.

Fibrinogen Degradation Product (FDP)

Latex Agglutination Test

Principle. Highly specific antibodies to fibrinogen degradation products D and E are adsorbed on latex particles, and the particles are suspended in glycine–saline buffer. Samples containing 2 or more than 2 μg FDP/ml agglutinate the latex particles when mixed together on a glass slide. The method is rather qualitative than quantitative.

Reagents

1. Thrombo Wellcotest Kit for FDP #HA13 (Wellcome Research). The kit contains the following:
   a. Latex suspension.
   b. Positive control serum.
   c. Negative control serum.
   d. Glycine–saline buffer.
   e. Specimen collection tubes.
   f. Glass slides.
   g. Disposable pipettes and mixing rods.

Procedure

A. Specimen collection.
   1. Draw blood into a syringe by clean venapuncture and draw 2 ml of blood into the specimen-collecting tubes provided with the kit. Each tube contains soybean trypsin inhibitor and bovine thrombin.
   2. Allow the blood to clot at 37°C for half an hour and collect clear serum after centrifugation at 2000 rpm for 15 minutes.
B. Positive control.
   1. Provided in Red Cap bottle.
   2. Contains human serum diluted in glycine–saline buffer containing 0.1% sodium azide to give an FDP concentration of 5–10 μg/ml.
C. Negative control.
   1. Provided in Blue Cap bottle.
   2. Contains diluted human serum to give an FDP concentration of less than 2 μg/ml.
D. Assay.
   1. Label two small tubes as 1 and 2.
   2. Add 0.75 ml glycine buffer into each tube. The graduated dropper from the bottle may be used.
   3. Using the disposable droppers provided with the kit, add five drops of the serum sample into tube 1 and one drop into tube 2.
   4. Mix the contents of each tube.
   5. Place one drop from tube 2 onto the glass slide at position 2 and one drop from tube 1 onto the slide at position 1.
   6. Mix the latex suspension and add one drop of the suspension to each position on the glass slide.
   7. Mix the serum–latex mixture with the wooden applicator and rock the slide gently for 2 minutes.
   8. Look for agglutination against a dark background.

Results

1. A positive agglutination indicates the presence of FDP in the serum. Since the sensitivity of the assay is adjusted to 2 μg FDP/ml, this amount should be multiplied by the dilution factor to determine the amount of FDP/ml of the sample.

Clinical significance

1. Fibrin(ogen) degradation products (FDP) are formed by the proteolytic action of plasmin on fibrin(ogen).
2. The fragments D (MW 90,000) and E (MW 50,000) are measured in order to assess primary and secondary fibrino(geno)lysis.
3. When present in high concentration, FDP may cause hemorrhage because of their antithrombin action at the site of local thrombus formation.

Plasminogen*

Radial Immunodiffusion (RID) Method

Principle. A monospecific antiserum to human plasminogen is incorporated into agarose and 12 small cylindrical wells are cut through the medium. Known concentrations of a standard plasma and patient samples are applied to the wells and immunodiffusion of the antigen is allowed to occur for 48 hours. The diameter of the precipitin ring is proportional to the amount of antigen in the sample. A standard curve of plasminogen concentration vs. the square of the ring diameter is plotted and the sample concentrations are read off the graph.

Materials required

1. M-Partigen Plasminogen Radial Immunodiffusion Plates (Calbiochem Behring Corp.). Store in refrigerator.
2. Protein standard plasma. Reconstitute with 0.5 ml distilled water. Good for 7 days after reconstitution.
3. Microliter pipette.
4. Measuring device, view box, or RID viewer.
5. Distilled water and normal saline.

---

* Source: Penner JA, Romond EH, Triplett DA: Antithrombin-III, Plasminogen and Fibrinogen in Thrombosis and Hemostasis. Wilmington, DuPont, 1982.

Figure 29. Measurement of plasminogen by a radial immunodiffusion method.

Specimen

1. Plasma, 100 μl, with citrate, EDTA, or oxalate.
2. Dilute specimen 1 : 2 with saline.

Procedure

1. Remove RID plates from the package and let them sit open at room temperature for 5 minutes.
2. Make 1 : 2 and 1 : 4 dilutions of the standard plasma and apply 5 μl of undiluted, 1 : 2, and 1 : 4 samples to wells 1, 2, and 3, respectively.
3. Apply 5 μl of patient(s) plasma or serum to wells 4–12.
4. Close the plate, replace aluminum cover, and keep at room temperature in a horizontal position.
5. Keep a record of the batch number of plates and the standard plasma, date and time of sample application, concentration of standards, and the patient number for each well.
6. Measure the diameter of the precipitin ring after 48 hours, using a calibrated measuring ruler and a view box. An RID viewer can also be used. Square the ring diameter reading (Fig. 29).
7. Plot a standard curve using concentrations of the three standards vs. squared ring diameters. Draw a straight line through the points which should intercept the Y-axis at 11 ± 3.5 mm.

8. Read the concentration of the patient sample from the standard curve and multiply by the dilution factor. Concentration values are obtained as mg/dl.

Normal range

1. Mean = 12.0 mg/dl.
2. Range = 10–20 mg/dl.

Nephelometric Immunoassay (NIA)

Principle. Antiserum to human plasminogen is added to disposable cuvettes. Reference sera and test specimen are then added and the cuvettes are incubated at room temperature for 1 hour. Following incubation, cuvettes are placed in a Hyland Laser Nephelometer PDQ and a beam of laser light is passed through the solution. The amount of light scattered by the antigen–antibody complexes is quantitatively measured within the range of the reference sera. The percentage of relative light scatter (% RLS) of the test sample is compared to a reference curve of RLS vs. plasminogen concentration in the reference sera.

Materials required

1. Las-R Human Plasminogen Test Kit (Hyland Diagnostics). The kit includes the following reagents:
   a. Plasminogen antiserum (goat).
   b. Antiserum diluent.
   c. Sample blank solution.
2. Las-R Reference Serum, Set 4.
3. Hyland Invertible Mixing Rack.
4. 10 × 75 mm borosilicate glass culture tubes (Scientific Products).
5. 10-ml plastic syringe; 0.45-μm Millipore filters (Millipore Corporation); Swin-Lok membrane holder (Nucleopore).
6. Normal saline.
7. Parafilm (American Can Company).

Reagent Preparation

1. Bring all reagents to room temperature before starting the test.
2. Make a 1 : 2 dilution of the plasminogen antiserum using antiserum diluent. Filter this solution using the 0.45-μm Millipore filter and the syringe. A single test requires 1 ml of antibody solution. Make as much antibody solution as required.

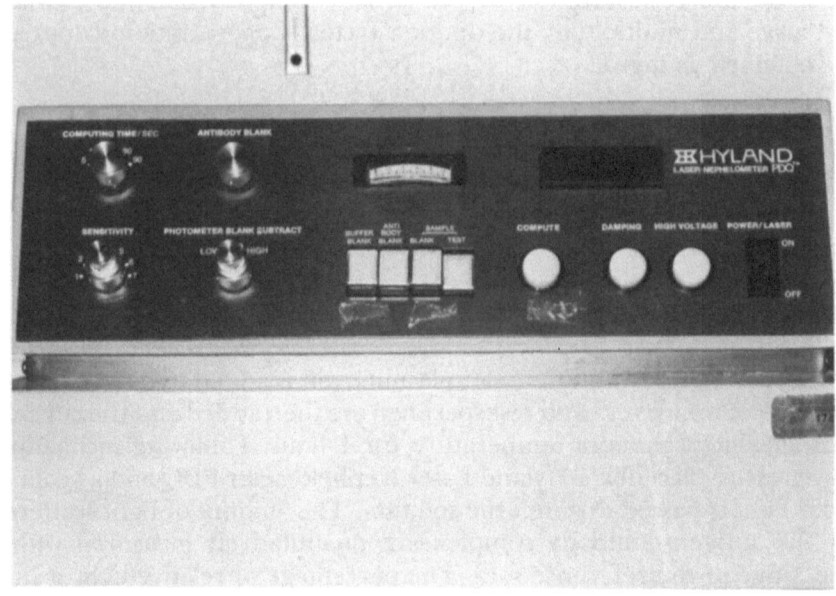

Figure 30. Hyland Laser Nephelometer PDQ, Courtesy of Hyland Diagnostics.

Instrumentation

1. Hyland Laser Nephelometer PDQ equipped with Hewlett Packard Microprocessor Module 9815 (Fig. 30).
2. Hyland Las-R Test Program #1 (Serial #62970-83).

Procedure

1. Arrange two rows of 10 × 75 mm borosilicate culture tubes in Hyland rack and label as 1, 2, 3, . . . *n*.
2. Dispense 1 ml sample blank solution into each tube in the first row.
3. Dispense 1 ml antibody solution into each tube in the second row.
4. Dispense 1 μl reference serum I into tube 1 in both first and second rows.
5. Dispense 1 μl reference serum II into tube 2 in both first and second rows. Dispense reference serum III, IV, V, and VI into tubes 3, 4, 5, and 6, respectively, in both first and second rows.
6. Dispense 1 μl patient(s) plasma into tubes 7, 8, . . . *n* in both first and second rows.
7. Cover the tubes with Parafilm, place the lid on top, and gently mix the solution by carefully inverting the rack.
8. Incubate the tubes at room temperature for exactly 1 hour.

Laser Nephelometer PDQ

1. Turn power switch on.
2. Set computing time to 5 seconds.
3. Set coarse sensitivity to 3 and fine sensitivity to 7.
4. Photomultiplier Blank Subtract knob should be at "medium" setting.
5. Sample Test button should be pushed in.
6. Push Compute button. The digital display should read zero; if not, adjust it to zero by using the zero set switch in the rear panel of the instrument.
7. At the end of the incubation period, turn on microprocessor and follow the instructions.

Normal range

1. Mean = 16 mg/dl.
2. Range = 8–23 mg/dl.

Clinical significance

1. Plasminogen is a glycoprotein with a molecular weight of 80,000 daltons. It is found in both plasma and serum.
2. It is activated by endothelial activator, kallikrein, streptokinase, and urokinase, resulting in the production of the proteolytic enzyme plasmin.
3. Increased levels of plasminogen have been reported in inflammation, bacterial infections, myocardial infarction, and after surgery.
4. Decreased levels of plasminogen are associated with thrombolytic therapy using urokinase and streptokinase, and in patients with DIC and other liver diseases.

Serine Protease Inhibitors

α_1-Antitrypsin (α_1-AT)*

Radial Immunodiffusion (RID) Method

Principle. The RID plates contain a mixture of agarose and monospecific antibodies to α_1-AT. Known concentrations of a standard serum are applied to a series of wells cut in the agarose. The antigen in sample

---

* Sources: Lewis JH, Iammarino RM, Spero JA, Hasiba U: Antithrombin Pittsburgh: an α_1-AT variant causing hemorrhagic disease. Blood 51:129–137, 1978; Messmore HL, Parvez Z, Fareed J. Isolation and partial characterization of a novel circulating antithrombin. Thromb Haemostasis 42(1):123, 1979.

diffuses into the medium and reacts with the antitrypsin to form a visible white precipitin ring. The area of the ring is proportional to the amount of antigen present in the sample. A standard curve with α_1-AT concentrations and the square of the ring diameters is plotted and the patient's sample concentration is read off the graph.

Materials required

1. M-Partigen α_1-AT Radial Immunodiffusion Plates (Calbiochem Behring Corp.). Store in refrigerator.
2. Protein standard plasma. Reconstitute with 0.5 ml distilled water. Good for 7 days after reconstitution when kept at 4°C.
3. Microliter pipette.
4. Measuring device, view box, or RID viewer.
5. Distilled water and normal saline.

Specimen

1. Plasma, 100 μl, citrate, EDTA, or oxalate.
2. Dilute specimen 1 : 10 with saline.

Procedure

1. Remove RID plate from the package and let it sit open, at room temperature, for 5 minutes.
2. Make 1 : 3, 1 : 6, and 1 : 12 dilutions of standard plasma in saline. Mix well and apply 5 μl of these dilutions to wells 1, 2, and 3, respectively.
3. Dilute patient(s) plasma. Apply 5 μl of patient(s) plasma to well(s) 4–12.
4. Close the plate, replace aluminum cover, and keep at room temperature in a horizontal position.
5. Keep a record of the batch number of plates and standard, date and time of sample application, concentration of standards, and patient number for each well.
6. Measure the diameter of the precipitin ring after 48 hours, using a calibrated measuring ruler and a view box. An RID viewer can also be used. Square the ring diameter readings (Fig. 31).
7. Plot a standard curve using concentrations of the three standards vs. squared ring diameters. Draw a straight line through the points which should intercept the Y-axis at 11 $\pm$ 3.5 mm.
8. Read the concentration of the patient samples from the standard curve and multiply by the dilution factor. Concentration values are obtained as mg/dl.

Normal range.   The normal range for human adults is 200–400 mg/dl.

Figure 31. Quantitation of α_1-antitrypsin by a radial immunodiffusion method.

Nephelometric Immunoassay (NIA)

Principle. A monospecific antiserum to human α_1-AT is mixed with a reference serum preparation in disposable cuvettes. After an incubation period of 1 hour at room temperature, the cuvettes are placed individually in a Hyland Laser Nephelometer PDQ, and a laser beam is passed through the cuvettes containing antigen–antibody solution. The amount of light scattered by the antigen–antibody complexes is measured as percent relative light scatter (% RLS), which is proportional to the antigen concentration in the solution. A six-point standard curve of reference serum concentration vs. % RLS is computed and the antigen concentration in unknown samples is obtained from this curve.

Materials required

1. Las-R Human α_1-Antitrypsin Test Kit (Hyland Diagnostics). The kit includes the following:
   a. α_1-AT antiserum (goat).
   b. Antiserum diluent.
   c. Sample blank solution.
2. Las-R Reference Serum, Set 2.
3. Hyland Invertible Mixing Rack.
4. 10 × 75 mm borosilicate glass culture tubes (Scientific Products).

5. 10-ml plastic syringe; 0.45-μm Millipore filters (Millipore Corporation); Swin-Lok membrane holder (Nucleopore).
6. Normal saline.
7. Parafilm (American Can Company).

Reagent preparation

1. Bring all solutions to room temperature before starting the test.
2. Make a 1 : 2 dilution of antiserum using antiserum diluent provided with the kit; filter this solution using a 0.45-μm Millipore filter and the syringe. A single test requires 1 ml of antibody. Make as much antibody solution as needed.

Instrumentation

1. Hyland Laser Nephelometer PDQ equipped with Hewlett Packard Microprocessor Module 9815 (Hyland Diagnostics).
2. Hyland Las-R Test Program #1 (Serial #62970-83).

Procedure

1. Arrange two rows of 10 × 75 mm borosilicate culture tubes in the Hyland rack and label them as 1, 2, 3, . . . n.
2. Dispense 1 ml of sample blank solution into each tube in the first row.
3. Dispense 1 ml of antibody solution into each tube in the second row.
4. Dispense 1 μl of reference serum I into tube 1 in both the first and second rows.
5. Dispense 1 μl of reference serum II into tube 2 in both the first and second rows and repeat the same procedure for tubes 3, 4, 5, and 6 in both the first and second rows, using reference serum III, IV, V, and VI, respectively.
6. Dispense 1 μl of patient(s) plasma into tubes 7, 8 . . . n in both the first and second rows.
7. Cover the tubes with Parafilm, and gently mix the solutions by carefully inverting the rack.
8. Incubate the tubes at room temperature for exactly 1 hour.

Laser nephelometer PDQ.   For a complete understanding of the operational procedure, follow Operator's Manual for the instrument.

1. Turn power switch on.
2. Set computing time to 5 seconds.
3. Set coarse sensitivity to 3 and fine sensitivity to 7.
4. Turn Antibody Blank knob fully clockwise. Damping button should be off.
5. Push High Voltage button.
6. Push Compute button.

7. Press Buffer Blank button and set Photomultiplier Blank Subtract knob to MED.
8. Press Sample Blank button.
9. Press Compute button. The digital display should read zero; if not, bring it to zero by using a Zero Set knob on the rear panel.
10. Push Sample Test knob.
11. At the end of the incubation period, turn on microprocessor.

Microprocessor module

1. Install the Las-R Test Program #1 Cassette Tape in the Hewlett Packard Calculator.
2. Printer Normal, Auto Start.
3. Turn power switch on and follow the instructions. A sample tape printout is given below:

AUTO START
PROGRAM 1
COMPUTE TIME IS 5
IF OK PRESS RUN
IF NOT PRESS NO
REF BLANKS?
YES
SET SENSITIVITY
COARSE 3X (Enter 3)
FINE 7X (Enter 7)
READ REF BLANKS
#1 = 9.9 RLS (Insert Ref. 1 and Press Compute)
#2 = 9.6 RLS (Insert Ref. 2 and Press Compute)
#3 = 12.0 RLS (Insert Ref. 3 and Press Compute)
#4 = 8.7 RLS (Insert Ref. 4 and Press Compute)
#5 = 7.1 RLS (Insert Ref. 5 and Press Compute)
#6 = 4.8 RLS (Insert Ref. 6 and Press Compute)
REF BLANKS STORED
HOW MANY SAMPLE BLANKS?
   5 (For example)
SET SENSITIVITY
COARSE 3X (Enter 3)
FINE 7X (Enter 7)
READ SAMPLE BLANKS
#1 = 8.0 RLS (Insert sample blank 1 and Compute)
#2 = 4.0 RLS (Repeat with 2, 3, 4, 5)
#3 = 7.0 RLS
#4 = 13.0 RLS
#5 = 17.0 RLS

SAMPLE BLANKS STORED
WHICH ASSAY?
IGA = 1
HAP = 3
TRA = 5
AAT = 7
SFG = 11
α_2-M = 8
CRP = 12
CPA = 15
T-1 = 17
T-3 = 19
C4 = 2
C3 = 4
ALB = 6
IGG = 10
IGM = 14
AT3 = 9
CER = 13
PMG = 16
T_2-2 = 18
T-4 = 20
ENTER AND PRESS RUN
7 (For α_1-AT)
INSERT R1 AND SET SENSITIVITY
COARSE SENSITIVITY?
 3 (Enter 3)
FINE SENSITIVITY?
 5 (Enter 5)
UNITS?
0 = MG/DL (HY)
1 = MG/DL (WHO)
2 = IU/ML (WHO)
 0 (Enter 0)
REFERENCE VALUES IN MEMORY
 747.0
 609.0
 465.0
 296.0
 149.0
  17.0
CORRECT?
YES

READ TEST REFERENCE CUVETTES
REF #1 = 180.0 RLS
REF #2 = 156.5 RLS
REF #3 = 130.6 RLS
REF #4 = 109.5 RLS
REF #5 = 77.0 RLS
REF #6 = 15.5 RLS

LIMITS
REF 1 CALL RLS 176
REF 6 CALL RLS
   17

R R = 0.9953
   AAT
OF UNKNOWNS?
   5 (Enter 5)
READ SAMPLES
#1 = 109.6 RLS
   306.8 MG/DL HY
#2 = 96.2 RLS
   242.1 MG/DL HY
#3 = 116.6 RLS
   343.6 MG/DL HY
#4 = 90.5 RLS
   216.8 MG/DL HY
#5 = 114.4 RLS
   331.9 MG/DL HY
REPEAT SAMPLE?
   NO
   0
NEW ASSAY?
   NO

Normal range

1. Mean = 248 mg/dl.
2. Range = 113–384 mg/dl.

Clinical significance

1. α_1-Antitrypsin is a primary inhibitor of trypsin, chymotrypsin, and elastase. It is also known to inhibit plasmin and thrombin. However, its exact role in coagulation physiology is not clearly known.

2. A genetic deficiency of α_1-AT results in emphysemalike pulmonary disease, whereas an acquired deficiency is known to cause lung destruction.
3. Variant forms of α_1-AT are associated with repeated bleeding episodes.

Antithrombin-III (AT-III)

Radial Immunodiffusion (RID) Method

Principle. A monospecific antiserum to human AT-III is incorporated into an agarose medium and 12 small cylindrical wells are cut through the medium. Known concentrations of a standard plasma and patient(s) sample(s) are applied to the wells and radial diffusion of the antigen is allowed to proceed for 48 hours. The diameter of the precipitin ring is proportional to the amount of antigen in the sample. A standard curve of AT-III concentrations vs. squared ring diameters is plotted and the concentration of the patient sample is read off the graph.

Materials required

1. M-Partigen AT-III Radial Immunodiffusion Plates (Calbiochem Behring Corp.). Store in refrigerator.
2. Protein standard plasma. Reconstitute with 0.5 ml distilled water. Good for 7 days after reconstitution when kept at 4°C.
3. Microliter pipette.
4. Measuring device, view box, or RID viewer.
5. Distilled water and normal saline.

Specimen

1. Plasma, 100 μl, citrated, EDTA, or oxalate.
2. Dilute specimen 1 : 2 with saline.

Procedure

1. Remove RID plate from the package and let it sit open at room temperature for 5 minutes.
2. Make 1 : 2 and 1 : 4 dilutions of standard plasma and apply 5 μl of undiluted, 1 : 2 and 1 : 4 samples to wells 1, 2, and 3.
3. Apply 5 μl of patient(s) plasma(s) to wells 4–12.
4. Close the plate, replace aluminum cover, and keep at room temperature in a horizontal position.

Figure 32. Determination of AT-III concentration in normal human plasma by a radial immunodiffusion method.

5. Keep a record of the batch number of plates and standard, date and time of sample application, concentration of standards, and the patient number for each well.
6. Measure the diameter of the precipitin ring after 48 hours, using a calibrated measuring ruler and a view box. An RID viewer can also be used. Square the ring diameter readings (Fig. 32).
7. Plot a standard curve using concentrations of the three standards and the squared ring diameters. Draw a straight line through the points which should intercept the Y-axis at 11 ± 3.5 mm.
8. Read the concentration of the patient sample from the standard curve and multiply by the dilution factor. Concentration values are obtained as mg/dl.

Normal range

1. Mean = 29.3.
2. Range = 18.8–39.8 mg/dl.

Nephelometric Immunoassay (NIA)

Principle. A monospecific antiserum to human AT-III is allowed to interact with reference serum preparations or patient plasma in disposa-

Table 15. Protocol for Antithrombin-III assay

Reactants	Blank Cuvettes (1 ml buffer)	Test Cuvettes (1 ml antiserum; 1 : 100)
NHP 200%	R_1	T_1
NHP 100%	R_2	T_2
NHP 50%	R_3	T_3
NHP 25%	R_4	T_4
NHP 6.25%	R_5	T_5
NHP 1.56%	R_6	T_6
Patient 1	P_1	P_1
Patient 2	P_2	P_2
Patient 3, etc.	P_3, etc.	P_3, etc.

R, blank; T, test; P, patient.
From Parvez et al: Thromb Res 24:367–377, 1981.

ble cuvettes. After an incubation period of 1 hour at room temperature, the cuvettes are placed individually in a Hyland Laser Nephelometer PDQ, and a laser beam is passed through the cuvette containing antigen–antibody solution. The amount of light scattered by the antigen–antibody complexes is measured as percent relative light scatter (% RLS), which is proportional to the antigen concentration in the solution. A six-point standard curve of reference serum concentration vs. % RLS is computed and the antigen concentration in unknown samples is obtained from this curve.

Materials required

1. Normal human plasma pool (NHP) is prepared from blood drawn from 20 healthy human volunteers (ten males and ten females). Collect blood in plastic tubes containing 3.8% sodium citrate in a 9 : 1 volume ratio of blood; citrate and centrifuge at 3000 rpm for 20 minutes. Pool all individual platelet-poor plasmas, make small aliquots, and store at −70°C for up to 3 months.
2. 0.05 M phosphate buffered saline (PBS), pH 7.4, containing 4% polyethylene glycol 6000 (PEG).
3. Antiserum to human AT-III raised in goats (Atlantic Antibodies).
4. 10 × 75 mm borosilicate glass culture tubes (Scientific Products).
5. 10-ml plastic syringe; 0.45 μm Millipore filters (Millipore Corporation); Swin-Lok membrane holder (Nucleopore).
6. Normal saline.

Reagent preparation

1. Make a 1 : 100 dilution of AT-III antiserum in PBS buffer and filter the solution by using a 0.45-μm Millipore filter. A single test requires 1 ml of antibody solution. Make as much antibody solution as needed.
2. Make serial dilutions of NHP using saline filtered through a Millipore filter and obtain 50, 25, 6.25, and 1.56% dilutions.
3. Centrifuge frozen patient samples at 3000 rpm for 15 minutes or filter through the Millipore filter, and make a 1 : 2 dilution by using filtered saline.

Instrumentation

1. Hyland Laser Nephelometer PDQ equipped with the Hewlett Packard Microprocessor Module 9815 (Hyland Diagnostics).
2. Hyland Las-R Test Program #1 (Serial #62970-83).

Procedure

1. Arrange two rows of 10 × 75 mm borosilicate tubes in a rack and label them as 1, 2, 3, etc.
2. Dispense 1 ml of PBS buffer into each tube in the first row.
3. Dispense 1 ml of diluted (1 : 100) antibody solution into each tube in the second row.
4. Dispense 2 μl of undiluted NHP (200%) into tube 1 in both first and second rows.
5. Dispense 1 μl of undiluted NHP (100%) into tube 2 in both first and second rows.
6. Dispense 1 μl of NHP (50%) into tube 3 in both first and second rows.
7. Dispense 1 μl of NHP (25%) into tube 4 in both first and second rows.
8. Dispense 1 μl of NHP (6.25%) into tube 5 in both first and second rows.
9. Dispense 1 μl of NHP (1.56%) into tube 6 in both first and second rows.
10. Dispense 1 μl of diluted (1 : 2) patient sample into tube 7 in both first and second rows.
11. Dispense 1 μl of diluted (1 : 2) patient samples into tubes 8, 9, 10 . . . n in both first and second rows.
12. Cover the tubes with Parafilm and mix the solutions by carefully inverting the rack.
13. Incubate the tubes at room temperature for exactly 1 hour.

Laser nephelometer PDQ

1. Turn power switch on. Set computing time to 5 seconds, coarse sensitivity to 3, and fine sensitivity to 7. Photomultiplier Blank subtract knob should be kept at "medium" level. Sample test button should be pushed in. Bring the display reading to zero, using the Zero Set switch in the rear panel of the instrument. Push Compute button to see if zero value is stabilized.
2. At the end of the incubation period, place tube 1 from the first row in the laser chamber, press Compute button, and record % RLS.
3. Place tube 2 from the first row in the laser chamber, press Compute button, and record % RLS.
4. Repeat the same procedure for all the tubes in the first row.
5. Place tube 1 from the second row in the laser chamber. Turn High Voltage button ON and, using the upper switch at the back of the instrument, bring the analogmeter pointer between 0.9 and 1 on the scale. Push Compute button and record % RLS.
6. Place tube 2 from the second row in the laser chamber and press Compute button. Do not adjust analogmeter setting. Record % RLS.
7. Repeat the same procedure with tubes 3, 4, 5, 6 . . . n.
8. Make a table of calibrator concentration vs. % RLS as follows:

Concentration	%RLS
200% (tube 1)	
100% (tube 2)	
50% (tube 3)	
25% (tube 4)	
6.125% (tube 5)	
1.56% (tube 6)	

9. Using linear graph paper, plot % RLS values on the vertical axis and concentration values on the horizontal axis, and draw a line of best fit. Read AT-III concentration in unknown samples from this graph.

Note: If the laser nephelometer is equipped with a Hewlett Packard Microprocessor Module 9815A, use test 9 in Hyland Las-R Test Program #1. Serial #62970-83 for calculations.

Normal range

1. 108 ± 13% (mean ± SD).

Rocket Immunoelectrophoresis

Principle.   A monospecific antiserum to AT-III is incorporated into an agarose medium and plasma containing AT-III is applied to a cylindrical well cut into the gel matrix. Upon electrophoresis, AT-III forms a rocket-shaped immunoprecipitation arc; the height of the rocket is proportional to AT-III concentration present in the plasma.

Materials required

1. Power supply (Helena).
2. Electrophoresis chamber (Helena).
3. Glass plates, 8 × 10 cm (Scientific Products).
4. Well cutters (Gelman).
5. Pasteur pipettes.
6. Microliter pipettes.

Reagents

1. Antiserum to human AT-III raised in rabbits (Calbiochem).
2. Seakem LE agarose (Marine Colloids).
3. Normal saline.
4. 0.05 M barbital buffer, pH 8.6.
   20.52 g sodium barbital (sodium diethylbarbiturate).
   3.68 g barbital (diethylbarbituric acid).
   2 liter distilled water.
5. Coomassie Brilliant Blue Stain (Kodak).
   Coomassie Brilliant Blue R-250, 0.50 g.
   Absolute methanol, 45.0 ml.
   Glacial acetic acid, 10.0 ml.
   Distilled water, 45.0 ml.
6. Destaining solution.
   Absolute methanol, 50.0 ml.
   Glacial acetic acid, 10.0 ml.
   Distilled water, 50.0 ml.

Specimen

1. Both frozen and freshly collected plasma samples can be used.
2. Dilute patient samples 1 : 2 using saline.
3. Make several dilutions of normal human pooled plasma (NHP) or any commercial reference plasma as follows:
   Undiluted = 100%
1 : 2	50%
1 : 4	25%
1 : 8	12.5%

Procedure

A. Precoating of glass plates.
   1. Wash the glass plates in distilled water and air dry.
   2. Precoat plates in 1% Seakem LE agarose made in distilled water.
   3. Let the plates air dry.
   4. When dry, mark the top with a marker pen.
B. Preparation of agarose plates.
   1. Dissolve 0.85 g Seakem agarose in 100 ml of 1:3 diluted buffer, and let it boil.
   2. Cool the agarose to 56°C and add 150 μl of anti–AT-III in 15 ml of agarose.
   3. Transfer 10–12 ml of antiserum–agarose mixture onto the precoated glass surface.
   4. Let the agarose settle at room temperature. Keep the plates covered in a refrigerator for at least 30 minutes after preparation.
   5. Using Gelman gel cutters, punch a series of 10 wells in the gel 2 cm from the top of the plate and 1 cm apart.
   6. Carefully remove agarose by suction through a Pasteur pipette.
   7. Apply 5 μl of NHP dilutions into wells 1–4, respectively, and patient samples into the remaining wells. The last well contains the tracking dye bromophenol blue.

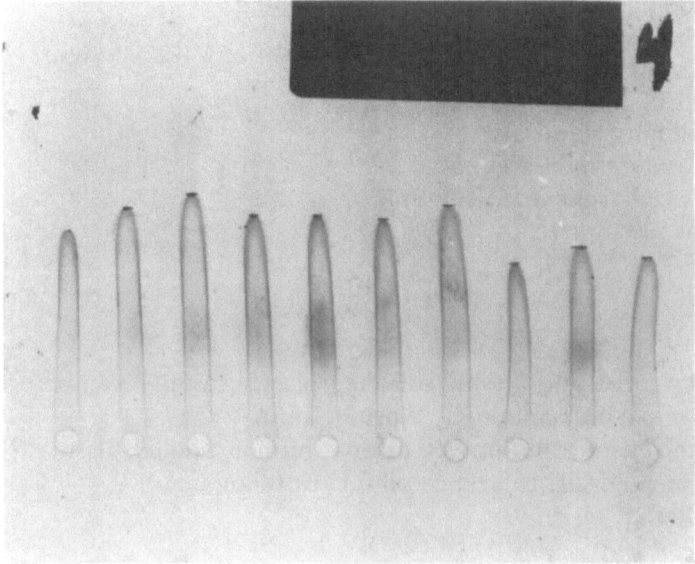

Figure 33. A rocket immunoassay for antithrombin-III in plasma.

Table 16. Rocket Immunoelectrophoresis Assay Protocol for
Antithrombin-III

Dissolve 0.85 g Seakem LE agarose in 100 ml of 1 : 3 diluted buffer
Buffer: 0.05 M barbital buffer, pH 8.6
Mix 150 μl of anti–AT-III in 15 ml agarose
Electrophoresis: 16 mA/plate, 8 × 10 cm, for 6 hours
Standard: NHP dilutions (1 : 1, 1 : 2, 1 : 4, 1 : 8)
Sample size: 5 μl/well patient sample diluted (1 : 2)
Stain: 0.50% Coomassie Blue
Make a standard curve of NHP dilutions vs. rocket height

---

8. Fill the chamber with 350 ml of buffer in each compartment.
9. Place the plate inverted onto the sponge wicks in the chamber, so that the wells are toward the cathode. Make sure that the plates are in good contact with the sponge wicks.
10. Run a current of 5 mA/plate for 20 minutes to concentrate the sample, and then increase it to 16 mA/plate for 6 hours.
11. At the end of electrophoresis wash the plates in saline for 12 hours, and then twice in distilled water for 1 hour each.
12. Dry the plates in an oven set at 60°C; keep a moist paper towel on top of the plates to ensure even drying.
13. Stain the dried plates in Coomassie Blue for 6–10 minutes.
14. Destain the plates in destaining solution.
15. Wash the plates in tap water to remove excess acetic acid.
16. Measure the distance from the top of the well to the top of the rocket peak (Fig. 33).
17. Plot a standard curve of plasma dilutions vs. rocket height and determine AT-III concentration in patient plasma from this curve.
18. Assay protocol is summarized in Table 16.

Clinical significance

1. Antithrombin-III is the principal physiologic inhibitor of thrombin and Factor Xa and is also known to inhibit Factors IXa, XIa, and XIIa.
2. AT-III concentration in normal plasma ranges from 18 to 30 mg/dl.
3. Decreased levels of AT-III are reported in acute venous thrombosis, liver disease, disseminated intravascular coagulation, in postoperative patients, and in women taking oral contraceptives.
4. A genetic deficiency of AT-III is associated with venous thromboembolic disease.

α_2-Macroglobulin (α_2-M)

Radial Immunodiffusion (RID) Method

Principle. A monospecific antiserum to human α_2-M is incorporated into an agarose medium and 12 small cylindrical wells are cut through the medium. Known concentrations of a standard plasma and patient samples are applied to the wells and radial diffusion of the antigen is allowed to proceed for 72 hours. The diameter of the precipitin ring is proportional to the amount of antigen in the sample. A standard curve of α_2-M concentration vs. squared ring diameters is plotted and the concentration of the patient sample is read off the graph.

Materials required

1. M-Partigen α_2-M Radial Immunodiffusion Plates (Calbiochem Behring Corp.). Store in refrigerator.
2. Prediluted Standard Serum B (solutions I, II, and III).
3. Microliter pipette.
4. Measuring device, view box, or RID viewer.
5. Distilled water and normal saline.

Specimen

1. Plasma, 100 μl, citrated, EDTA, or oxalate.
2. Dilute specimen 1 : 2 with saline.

Procedure

1. Remove RID plate from the package and let it sit open at room temperature for 5 minutes.
2. Apply 5 μl of solutions I, II, and III to wells 1, 2, and 3, respectively.
3. Apply 5 μl of patient(s) plasma or serum to wells 4–12.
4. Close the plate, replace aluminum cover, and keep at room temperature in a horizontal position.
5. Keep a record of the batch number of plates and standards, date and time of sample application, concentration of standards, and the patient number for each well.
6. Measure the diameter of the precipitin ring after 72 hours, using a calibrated measuring ruler and a view box. An RID viewer can also be used. Square the ring diameter reading (Fig. 34).
7. Plot a standard curve using concentrations of the three standards vs. squared ring diameters. Draw a straight line through the points which should intercept the Y-axis at 11 ± 3.5 mm.

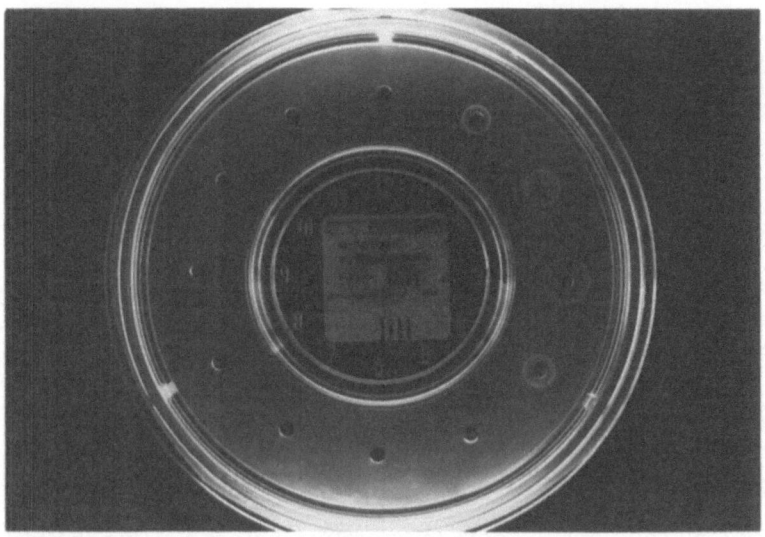

Figure 34. Measurement of α_2-macroglobulin in plasma by a radial immunodiffusion method.

8. Read the concentration of the patient sample from the standard curve and multiply by the dilution factor. Concentration values are obtained as mg/dl.

Normal range

1. The normal range in adult males is 150–350 mg/dl and 175–420 mg/dl in adult females.

Nephelometric Immunoassay (NIA)

Principle.   A beam of laser light is passed through an antigen–antibody solution and the light scattered by the antigen–antibody complexes is quantitatively measured within the range of the reference sera. The amount of light scattered is proportional to the concentration of the antigen. This light scatter is measured as percent relative light scatter (% RLS). The RLS of the unknown samples is compared to a reference curve constructed from the RLS of the reference sera and assigned α_2-M values.

Materials required

1. Las-R-Human α_2-Macroglobulin Test Kit (Hyland Diagnostics). The kit includes the following reagents:
   a. α_2-Macroglobulin antiserum (goat).
   b. Antiserum diluent.
   c. Sample blank solution.
2. Las-R Reference Serum, Set 4.
3. Hyland Invertible Mixing Rack.
4. 10 × 75 mm borosilicate glass culture tubes (Scientific Products).
5. 10-ml plastic syringe; 0.45-μm Millipore filters (Millipore Corporation); Swin-Lok membrane holder (Nucleopore).
6. Normal saline.
7. Parafilm (American Can Company).

Reagent preparation

1. Bring all solutions to room temperature before starting the test.
2. Make a 1 : 2 dilution of antiserum using antiserum diluent provided with the kit. Filter this solution using a 0.45-μm Millipore filter and the syringe. A single test requires 1 ml of antibody solution. Make as much antibody solution as needed.

Instrumentation

1. Hyland Laser Nephelometer PDQ equipped with Hewlett Packard Microprocessor Module 9815 (Hyland Diagnostics).
2. Hyland Las-R Test Program #1 (Serial #62970-83).

Procedure

1. Arrange two rows of 10 × 75 mm borosilicate culture tubes in a Hyland rack and label as 1, 2, 3, . . . n.
2. Dispense 1 ml sample blank solution into each tube in the first row.
3. Dispense 1 ml of antibody solution into each tube in the second row.
4. Dispense 1 μl reference serum I into tube 1 in both first and second rows.
5. Dispense 1 μl reference serum II into tube 2 in both first and second rows. Dispense reference serum III, IV, V, and VI in tubes 3, 4, 5, and 6 in both first and second rows.
6. Dispense 1 μl patient(s) plasma into tubes 7, 8 . . . n in both first and second rows.
7. Cover the tubes with Parafilm, place the lid on the top of the rack, and gently mix the solution by carefully inverting the rack.
8. Incubate the tubes at room temperature for exactly 1 hour.

Laser nephelometer PDQ

1. Turn power switch on.
2. Set computing time to 5 seconds.
3. Set coarse sensitivity to 3 and fine sensitivity to 7.
4. Photomultiplier Blank Subtract knob should be at "Medium."
5. Sample Test button should be pushed in.
6. Push Compute button. The digital display should read zero. If not, bring it to zero by using a Zero Set switch in the rear panel of the instrument.
7. At the end of the incubation period, turn on the microprocessor and follow the instructions as detailed in the α_1-AT procedure.

Normal range

1. Mean = 194 mg/dl.
2. Range = 107–281 mg/dl.

Clinical significance

1. α_2-Macroglobulin is a broad spectrum inhibitor of serine proteases and inhibits plasmin, kallikrein, and thrombin.
2. α_2-M concentration is increased during pregnancy and in women taking oral contraceptives.
3. Decreased levels of α_2-M are seen during the infusion of plasminogen activators such as streptokinase and urokinase.

α_2-Antiplasmin (α_2-AP)

Rocket Immunoelectrophoresis

Principle. The general principle is the same as that described for Factor VIII : RAg. Assay protocol is shown in Table 17.

Materials required

1. Power supply (Helena).
2. Electrophoresis chamber (Helena).
3. Glass plates, 8 × 10 cm (Scientific Products).
4. Well cutters (Gelman).
5. Pasteur pipettes.
6. Microliter Pipettes.

Table 17. Rocket Immunoelectrophoresis Assay
Protocol for α_2-Antiplasmin

0.85% Seakem LE agarose in diluted (1 : 3) buffer
150 μl anti–α_2-AP/15 ml agarose
0.05 M barbital buffer, pH 8.6
Electrophoresis: 16 mA/plate, 8.3 × 10.2 cm, for 6 hours
Standard: NHP dilutions (1 : 8, 1 : 16, 1 : 32, 1 : 64)
Sample size: 5 μl/well
Stain: 0.50% Coomassie Blue
Prepare a standard curve of NHP dilutions vs. rocket height

Reagents

1. Antiserum to human α_2-antiplasmin (Nordic).
2. Seakem LE Agarose (Marine Colloids).
3. Normal saline.
4. 0.05 M barbital buffer, pH 8.6.
    20.52 g sodium barbital (sodium diethylbarbiturate).
    3.68 g barbital (diethylbarbituric acid).
    2 liter distilled water.
5. Coomassie Brilliant Blue Stain (Kodak).
    Coomassie Brilliant Blue R-250, 0.50 g.
    Absolute methanol, 45.0 ml.
    Glacial acetic acid, 10.0 ml.
    Distilled water, 45.0 ml.
6. Destaining solution.
    Absolute methanol, 50 ml.
    Glacial acetic acid, 10 ml.
    Distilled water, 50 ml.
7. Tracking dye, bromophenol blue.

Specimen

1. Both frozen and freshly collected plasma samples can be used.
2. Dilute patient plasma 1 : 8 and 1 : 16, using diluted (1 : 3) barbital buffer.
3. Make serial dilutions of normal human pooled plasma (NHP) or any commercial reference plasma using diluted (1 : 3) barbital buffer.
4. Use 1 : 8, 1 : 16, 1 : 32, and 1 : 64 dilutions of the reference plasma or NHP to construct a standard curve.

Procedure

A. Precoating of glass plates.
   1. Wash the glass plates in distilled water and air dry.
   2. Precoat glass plates with 1% Seakem LE agarose dissolved in distilled water.
   3. Let the plates air dry.
   4. When dry, mark the top of the plates with a felt pen.
B. Preparation of antiserum–agarose plates.
   1. Dissolve 0.85 g Seakem LE agarose in 100 ml of 1 : 3 barbital buffer and let it boil.
   2. Cool the agarose to 56°C and add 150 μl of undiluted α_2-antiplasmin antiserum in 15 ml of melted agarose.
   3. Using a prewarmed glass pipette, transfer 10 ml of the above agarose–antiserum mixture onto a precoated glass plate.
   4. Let the agarose set at room temperature. Keep the plates covered in a refrigerator for at least 30 minutes after preparation.
   5. Using Gelman well cutters, punch a series of 10 wells in the gel 2 cm from the top of the plates and 0.5–0.8 cm apart.
   6. Remove agarose from the wells by suction through a Pasteur pipette without breaking the edge of the wells.
   7. Patient samples are used at dilutions of 1 : 8 and 1 : 16.
   8. Apply 5 μl of reference plasma or NHP dilutions into wells 1–4, respectively, and patient samples into the remaining wells. The last well contains the tracking dye, bromophenol blue.
   9. Dispense barbital buffer into electrophoresis chamber.
   10. Place plates in the chamber, so that the wells are toward the cathode.
   11. Press the plates on the sponge wicks to ensure complete contact with the agarose. (The plates have to be turned upside down.)
   12. Electrophorese at 16 mA/plate for 6 hours.
   13. At the end of electrophoresis, wash plates in saline for 12 hours.
   14. Wash plates twice in distilled water for 30 minutes each.
   15. Dry the plates in an oven set at 60°C; keep a moist filter paper on top of the plates to ensure even drying.
   16. Stain the dried plates in Coomassie Blue for 6–10 minutes.
   17. Destain the plates in destaining solution.
   18. Wash the plates in tap water to remove excess acetic acid.
   19. Measure the distance from the top of the well to the top of the peak.
   20. Plot a standard curve (reference plasma dilutions) on linear graph paper using the X-axis for percent α_2-AP and the Y-axis for rocket height in centimeters (Fig. 35).

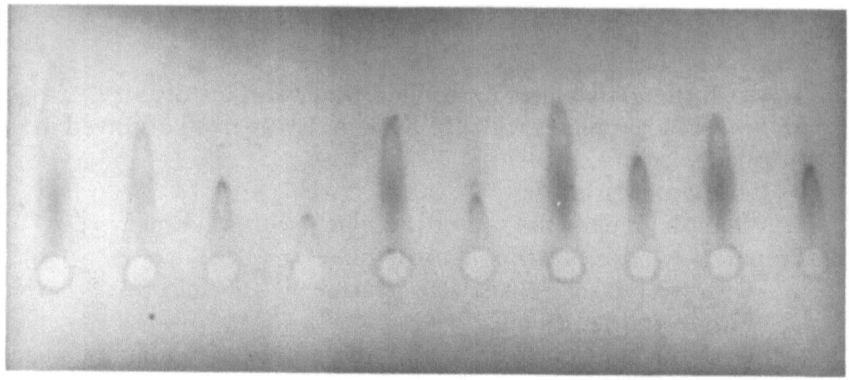

Figure 35. A rocket immunoassay for measuring α_2-antiplasmin in human plasma.

21. Determine the α_2-AP concentration in patient samples from the standard curve.

Normal range

1. 84–112% of normal. Normal is a plasma pooled from 10 male and 10 female healthy volunteers and is assumed to contain 100% of α_2-AP.

Clinical significance

1. α_2-Antiplasmin is a fast-acting inhibitor of plasmin, and is also known to inhibit Factor XII fragments, kallikrein, Factor IX, and thrombin. It may also inhibit urokinase and Factor Xa.
2. Decreased levels of α_2-AP are seen in liver diseases, disseminated intravascular coagulation, and during thrombolytic therapy.
3. A genetic deficiency of α_2-AP has been reported in four families; a homozygous state is responsible for severe bleeding in such patients.

Platelet Specific Proteins

Platelet Factor 4 (PF4)

Radioimmunoassay (RIA)

Principle. A constant amount of ^{125}I-labeled PF4 is mixed with PF4 antiserum and the nonradioactive PF4 in plasma is allowed to compete for the binding sites on the antiserum. The percentage of labeled PF4 bound to antiserum is inversely proportional to the amount of nonlabeled PF4 in the patient plasma. The bound antiserum is separated from the unbound by precipitation with ammonium sulfate, and the radioactivity in the bound antiserum complex is measured with a well-type gamma scintillation counter. PF4 concentration in the test plasma is determined from a standard curve of nonradioactive PF4 concentration vs. percent bound antiserum.

Materials required

1. Plastic test tubes, 12 × 75 mm, and rack.
2. 7 cc liquid EDTA vacutainer tubes and vacutainer needle.
3. Microliter pipettes (50, 250, and 500 μl).
4. Absorbent paper.
5. Vortex mixer.
6. Ice bath.
7. 1-ml pipetting syringe.
8. Timer.
9. Biogamma Counting Vials.
10. Biogamma Counter and Calculator/Printer (Beckman Instruments).
11. Linear graph paper.

Reagents

1. PF4 RIA Diagnostic Kit (Abbott Laboratories). This kit contains the following:
    a. ^{125}I-Platelet factor 4 (human).
    b. Platelet factor 4 antiserum (goat).
    c. Dilution buffer.
    d. Platelet factor 4 (human) standards.
    e. Ammonium sulfate.

Procedure

A. Specimen Collection
    1. Prepare a mixture of crushed ice and water.
    2. Using a 19- or 20-gauge needle and the vacutainer, draw two samples of blood into EDTA tubes. Discard the first tube to avoid tissue thromboplastin; after the second tube is filled, mix the blood and within 10 seconds place the tube on the melting ice bath.
    3. After 30 minutes on ice, remove the rubber stopper and centrifuge at 5000 rpm at 4°C for 20 minutes.

Table 18. Radioimmunoassay Procedure for Platelet Factor 4

Add 50 μl patient plasma to a plastic tube

Add 250 μl ^{125}I-PF4

Add 250 μl antiserum; mix and incubate for 1 hour at room temperature

Add 1 ml ammonium sulfate and mix

Centrifuge at room temperature for 20 minutes (300 rpm)

Decant and count precipitated antigen–antibody complex

Plot standard curve and read sample value

Compiled from Abbott Laboratories insert.

4. Aspirate 0.5 ml of plasma from 1 cm below the liquid surface of the tube and transfer it to a plastic tube. The sample can be stored in a refrigerator for 24 hours or at $-70°C$ for up to 3 months.

B. Assay (a summary of the assay procedure is presented in Table 18).
1. Bring all test kit reagents and specimens to room temperature before beginning test.
2. Label tubes as follows:
   - 1, 2, 3: total count tubes (TCT)
   - 4, 5:   nonspecific binding (NSB)
   - 6, 7:   zero standard
   - 8, 9:   10 ng/ml standard
   - 10, 11: 30 ng/ml standard
   - 12, 13: 50 ng/ml standard
   - 14, 15: 100 ng/ml standard
   - 16, etc.: for patient specimens and controls, in duplicate
3. Add 50 μl of dilution buffer into tubes 4–5 (NSB) and 6–7 (zero standard). Pipette 50 μl of standards and patient samples into appropriate tubes.
4. Add 250 μl of ^{125}I-PF4 reagent solution into all tubes. Cap tubes 1–3 and set aside.
5. Pipette 250 μl of dilution buffer into tubes 4–5 (NSB) and pipette 250 μl of PF4 antiserum into tubes beginning with 6.
6. Mix all tubes on the vortex for 3–5 seconds, and incubate at room temperature for 2 hours.
7. Pipette 1 ml of ammonium sulfate solution into all tubes except 1–3. Vortex for 3–5 seconds, then allow to stand 60 minutes.
8. Centrifuge all tubes except 1–3 at room temperature at 3000 rpm for 20 minutes.
9. Place tubes into a rack, and decant the supernatant very carefully by inverting the rack of tubes over the sink with the water running. Allow rack to rest inverted on tissues for a few mo-

ments, then gently blot rim of tubes with tissues or absorbent paper towels.

10. Place tubes in Biogamma counter and read the net radioactivity in counts per minute (cpm).
    a. Power: on (always).
    b. Count light: off (Mode A).
    c. Time: 1 minute.
    d. Low Count: off.
    e. Mode: manual.
    f. Push toggle switch on right rear of drawer to line up vial carriers.
    g. Push Reset, a square red button behind the orange panel, to 1.
    h. On computer terminal, put toggle to "online," and put power on.
    i. Change mode switch to "automatic" and let run.

C. Calculation.

1. $$\frac{\text{cpm of standard/control/unknown}}{\text{average cpm of TCT}} \times 100$$

$$= \% \text{ bound PF4 for standard or unknown}$$

2. Plot the average percent bound PF4 for each PF4 standard on the vertical axis vs. the known concentration in ng/ml on the horizontal axis. Draw the best-fit smooth curve.

3. Read the control and unknown concentrations from the graph.

D. Limitations.

1. Plasma containing high lipid levels should be run undiluted, $1:2$, and $1:4$. High lipid levels may cause nonlinearity, in which case the corrected levels obtained at $1:2$ and $1:4$ dilutions will be more accurate.

2. The lowest level of sensitivity as reported by the manufacturer is 2.5 ng/ml, although in our lab we have obtained lower values on normal donors and patients.

Normal range

1. Mean = 4.4 ng/ml.
2. Range = 0–10.4 ng/ml.

Clinical significance

1. Platelet factor 4 is a platelet specific protein secreted by the alpha-granules during the release reaction.
2. Its presence in plasma indicates platelet activation, which may occur in several pathophysiologic disorders.
3. Elevated levels of PF4 in plasma have been reported in coronary artery disease, myocardial infarction, and diabetes.

Beta-Thromboglobulin (β-TG)

Radioimmunoassay (RIA)

Principle.   The assay is based on the competition between unlabeled β-TG in the plasma sample and radioactively labeled β-TG for binding sites on a β-TG–specific antibody. The ^{125}I-β-TG–antibody complex is separated from unbound ^{125}I-β-TG by precipitation with ammonium sulfate. The precipitate containing the β-TG–antibody complex is then separated by centrifugation and its radioactivity measured in a gamma counter. The amount of radioactivity in the precipitate indicates the amount of ^{125}I-β-TG present, and is inversely proportional to the β-TG level in the plasma.

Materials required

1. Gamma Scintillation Counter (Beckman).
2. Refrigerated centrifuge capable of 10,000 rpm.
3. Microliter pipettes (50, 200, and 500 μl).
4. 12 × 75 mm polystyrene test tubes, rounded bottoms.
5. Vortex mixer.
6. Decantation racks.
7. Test tube racks.
8. Parafilm.

Reagents

1. β-TG RIA Kit (Amersham). This kit contains the following:
   a. Standard β-TG (human) in buffer (freeze-dried). Reconstitute each standard vial with 0.5 ml distilled water. Allow the standard to dissolve at room temperature for 2 minutes.
   b. ^{125}I-β-TG (human; freeze-dried). Contains up to 2 μCi ^{125}I per vial. Reconstitute vial with 10 ml distilled water.
   c. Anti–β-TG (rabbit; freeze-dried). Contains antiserum sufficient to bind at least 40% of 0.5 ng β-TG. Reconstitute with 10 ml distilled water.
   d. Ammonium sulfate solution. Contains a 3.3 M solution of ammonium sulfate.
   e. Blood collection tubes. Contain anticoagulant and antiplatelet agents.

Sample preparation

1. Place blood collection tubes provided with the kit in a crushed-ice water bath.

2. Using a 20-gauge × 1″ needle collect 5–10 ml blood in a polystyrene syringe. A total venipuncture time of less than 2 minutes is recommended.
3. Remove needle from the syringe and gently add 2.5 ml of blood to the precooled, labeled sampling tube.
4. Replace cap on tube, gently invert the tube to mix the blood with the anticoagulant, and immediately place blood specimen back in the cooling bath.
5. Cool the blood sample for approximately 30 minutes and centrifuge at 1500–2000 × g and 2–4°C for 30 minutes.
6. After centrifugation, carefully transfer the top 0.5 ml of plasma to a labeled polystyrene test tube. The sample may be stored at room temperature for up to 24 hours, or up to 4 weeks at −20°C.

Assay

1. Label 12 × 75 mm test tubes as 1–10 for the standards and two tubes for each patient.
2. Working from the lowest standard to the highest and starting with tube 1, place 50 μl of reconstituted standard into two test tubes (test will be done in duplicate). For example, place 50 μl of lowest standard in tubes 1 and 2, 50 μl of next standard in tubes 3 and 4, and so on.
3. Starting with tube 11, place 50-μl aliquots of patient plasma into duplicate tubes. For example, place 50 μl of plasma from patient #1 into tubes 11 and 12, 50 μl of plasma from patient #2 into tubes 13 and 14, and so on.
4. Pipette 200 μl of the ^{125}I-β-TG solution into each assay tube. Vortex for 3–5 seconds.
5. Pipette 200 μl of the anti–β-TG serum solution into each assay tube. Vortex for 3–5 seconds.
6. Incubate the assay tubes at room temperature (15–30°C) for 1 hour (±10 minutes). Place Parafilm over the rack of tubes during this incubation.
7. Pipette 500 μl of ammonium sulfate solution into each assay tube and immediately vortex each tube thoroughly. Efficient mixing is essential and this step should be performed within 8 minutes.
8. Centrifuge tubes for 10–15 minutes at 1000–1500 × g, at room temperature.
9. Place tubes in decantation racks and carefully pour off supernatants so as not to disturb the precipitate at the bottom of the tubes.
10. Keeping the tubes inverted, place them on paper tissues and drain for 5 minutes.

11. Gently touch the rims of the tubes with paper tissue to remove any remaining liquid before returning tubes to an upright position.
12. Count the radioactivity in the precipitates in a gamma counter. Count for either 1 or 2 minutes; calculate results as counts per minute (cpm).

Results

1. Using linear graph paper, plot a curve of ^{125}I counts for the five standards vs. the concentration stated on each vial. Plot the duplicate points for each standard and draw a smooth curve through the means of the duplicate points.
2. Using the mean of the duplicate points for each patient sample assayed, determine the β-TG level from the standard curve.

Normal range

1. Mean = 35 ng/ml.
2. Range = 13–40 ng/ml.

Clinical significance

1. Beta-thromboglobulin is a platelet specific protein secreted by the alpha-granules during platelet activation.
2. Although no specific function has been assigned to β-TG, its excessive release into the circulation acts as a marker for platelet release reaction.
3. Elevated levels of β-TG have been reported in hypercoagulable states and in post–cardiac surgery patients.

Thromboxane B_2 (TxB$_2$)

*Radioimmunoassay**

Principle. Unlabeled thromboxane B_2 from standard or test sample is allowed to react with a fixed amount of labeled TxB$_2$ and a limited amount of antibody. Unlabeled TxB$_2$ will compete with the labeled antigen for antibody binding sites. The unbound antigen is separated from the antigen–antibody complexes and radioactivity in the complexes is counted in a gamma counter. A plot of percent B/B$_0$ for each standard

---

* This radioimmunoassay is used for research purposes only. TxB$_2$ assay kit (Catalog No. 6221) is also available from Seragen, Boston, MA.

vs. the amounts of TxB_2 added is constructed and the concentration of TxB_2 in test samples is determined from this curve.

Materials required

1. Siliconized glass or polypropylene test tubes.
2. Microliter pipettes.
3. Ice bath.
4. Beakers.
5. Test tube rack.
6. Vortex mixer.
7. Refrigerated centrifuge.
8. Gamma counter.
9. Linear graph paper.

Reagents

1. Thromboxane B_2 ^{125}I-RIA Kit (Catalog No. Nek-024, New England Nuclear). The kit contains the following:
   a. Antiserum to TxB_2.
   b. TxB_2 ^{125}I-tracer concentrate.
   c. TxB_2 standard, lyophilized.
   d. Prostaglandin-free human plasma.
   e. Assay buffer.
   f. Precipitating solution.
   Store the kit at 4°C.

*Sample preparation**

1. Collect the blood in prechilled siliconized glass tubes containing 4.5 mM EDTA and indomethacin (10 μg/ml).
2. Centrifuge blood at 3000 rpm for 15 minutes and collect plasma. Plasma can be stored at -70°C if it is not assayed immediately.

Procedure

1. Reconstitute the TxB_2 standard with 1.0 ml of distilled water (yielding 100 ng/ml).
2. Using the assay buffer prepare a series of working standard solutions:

---

* A prior extraction of prostaglandins from plasma is also recommended.

Tube		Concentration (pg/0.1 ml)
A.	0.1 ml standard + 1.9 ml assay buffer	500
B.	0.2 ml A + 0.8 ml assay buffer	100
C.	0.4 ml B + 0.4 ml assay buffer	50
D.	0.4 ml C + 0.4 ml assay buffer	25
E.	0.4 ml D + 0.6 ml assay buffer	10
F.	0.4 ml E + 0.4 ml assay buffer	5
G.	0.4 ml F + 0.4 ml assay buffer	2.5
H.	0.4 ml G + 0.6 ml assay buffer	1

3. Add 100 μl of normal human plasma to all standard solutions.

Assay

1. Assay protocol is shown in Table 19.
2. After mixing the solutions, incubate tubes for 16 hours at 4°C.
3. After incubation, place tubes in an ice bath.
4. Add 1.0 ml of cold precipitating solution into all tubes beginning with tube 3. Vortex each tube for 2–5 seconds.
5. Incubate tubes at 2–8°C for another 30 minutes.
6. Centrifuge the tubes in a refrigerated centrifuge at 3000 rpm for 30 minutes.
7. Decant the supernatant of all tubes beginning with tube 3 and drain on absorbent paper. Blot the tubes to remove moisture.
8. Count all tubes in a gamma counter for 1 minute.
9. Calculate results.

Calculations

1. Average the counts for each set of duplicates.
2. Calculate the average *net* counts by subtracting average blank counts from each standard and sample.

Table 19. Assay Protocol for Thromboxane B_2

Solution	Tube No.	Buffer	PG-Free Plasma	Standard	Sample	Tracer	Antibody
Total counts	1–2	–	–	–	–	100	–
Blank	3–4	200	100	–	–	100	–
0 standard	5–6	100	100	–	–	100	100
Standards	7–20	–	100	100	–	100	100
Samples	21, 22, etc.	100	–	–	100	100	100

All volumes are in microliters.

3. Determine percent bound ($\% B/B_0$) for each standard and sample:

$$\% B/B_0 = \frac{\text{Net CPM of standard or sample}}{\text{Net CPM of 0 standard}} \times 100$$

4. Plot $\% B/B_0$ for each standard vs. the amounts of TxB_2 added (in pg) and determine TxB_2 concentration in test samples by interpolation from the standard curve.

Clinical significance

1. In platelets, endoperoxides PGG_2 and PGH_2 are transformed mainly into thromboxane A_2 ($t_{1/2}$ = 32 seconds) and then to thromboxane B_2, which is a stable product.
2. Stimulated platelets produce thromboxane A_2, which aggregates other platelets and causes constriction of arterial smooth muscles.
3. Production of TxA_2 can be implicated in the pathogenesis of thrombosis.
4. In diabetes, coronary thrombosis, and certain hyperlipidemias overproduction of TxA_2 has been reported.

Suggested readings

de Boer AC, Genton E, Turpie AGG: Chemistry, measurement, and clinical significance of platelet specific proteins. Crit Rev Clin Lab Sci 18(2):183–211, 1982.

New England Nuclear. Thromboxane B_2, ^{125}I-RIA Kit. Instruction Manual, 1–19, 1982.

6

Immunotechniques in Coagulation Research

Immunologic techniques have played a key role in both basic and applied research in coagulation. Besides providing total immunoreactive concentration of a coagulation protein, these methods have contributed a great deal to our understanding of the molecular abnormalities in fibrinogen, prothrombin, and other vitamin K–dependent coagulation factors,[24,38,41,66,72,74,76,89,90,104] as well as provided new avenues of diagnosis between hemophilia and von Willebrand's disease.[79] Highly sensitive assays such as radioimmunoassay, immunoradiometric assay, radioimmunoelectrophoresis, and crossed radioimmunoelectrophoresis[19,36,47,123] have enabled us to detect as low as 0.1% concentration of Factor VIII : RAg.

In addition to these quantitative uses, crossed immunoelectrophoresis and enzyme-linked immunosorbent assays are being utilized to detect molecular complexes between proteases and their inhibitors (e.g., thrombin : antithrombin-III, plasmin : α_2-antiplasmin) as seen in consumptive coagulopathies, septicemia, and malignancies. In this chapter some aspects of the use of immunoassays in coagulation research will be discussed.

Double Immunodiffusion

Double immunodiffusion or Ouchterlony technique can be utilized in the identification of coagulation proteins. In this technique, a set of wells is cut around a central well in an agarose medium. Experimental samples from protein purification steps are placed in the outer wells and a monospecific antiserum is incorporated into the central well. For example, in the purification of antithrombin-III from an affinity column, the separated fractions can be tested against antibodies to albumin, antithrombin-III, and α_1-antitrypsin. The fractions showing a single precipitin arc only with antibodies to antithrombin-III are pooled and taken as purified antithrombin-III. The purity of the material can be further tested in an immunoelectrophoresis system.

Crossed Immunoelectrophoresis

Both immuno- and crossed immunoelectrophoresis are qualitative techniques capable of detecting microheterogeneity in proteins differing in only 1 or 2 amino acids and in 0.02 units of their isoelectric points. These methods have been extensively utilized in studying variant forms of fibrinogen, prothrombin, and Factors X, VII, VIII, and IX. Abnormalities of Factor VIII/vWF in severe von Willebrand's disease (vWd) have been established by using crossed immunoelectrophoresis and SDS-agarose electrophoresis. Based upon electrophoretic migration of Factor VIII/vWF, types of vWd have been recognized. In vWd type I, levels of Factor VIII : C, VIII : RAg, and VIII : RCO are decreased to the same extent. On crossed immunoelectrophoresis, VIII : RAg shows slow- and fast-moving (multimeric) forms. In vWd type II, levels of these coagulant and antigenic proteins are not equally decreased. In crossed immu-

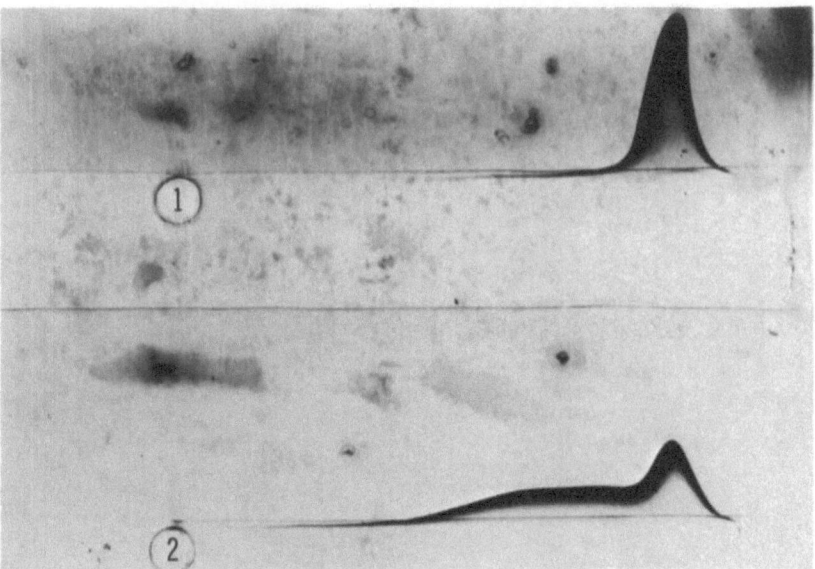

Figure 36. Detection of α_1-AT–protease complexes in leukemic cell lysate by a crossed immunoelectrophoresis technique. **1** Free α_1-AT. **2** Complexed α_1-AT showing slow mobility in leukemic cell lysate.

noelectrophoresis and SDS-agarose electrophoresis, the large slow-moving form is not present and an increased concentration of smaller multimers is indicated. The technique of crossed immunoelectrophoresis is also employed to investigate the consumption of serine protease inhibitors and other coagulation factors in disseminated intravascular coagulation, septicemia, and leukemic states.[28,35,85] In vitro studies from our laboratories have shown complexes between proteases liberated from leukemic cells and serine protease inhibitors (Figs. 36 and 37).

Antibody Neutralization Assay

Antibody neutralization was applied in 1956 to study inherited coagulation defects.[31] Allogenic antibodies (inhibitors) develop in patients with

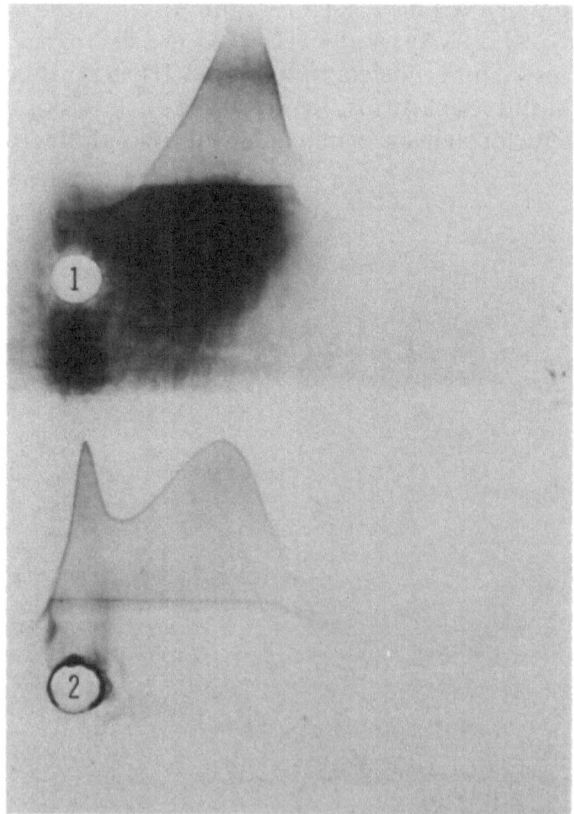

Figure 37. Complex formation between α_2-macroglobulin and the proteases from leukemic cell lysate. **1** Free α_2-macroglobulin. **2** Complexed α_2-macroglobulin showing slow mobility in leukemic cell lysate.

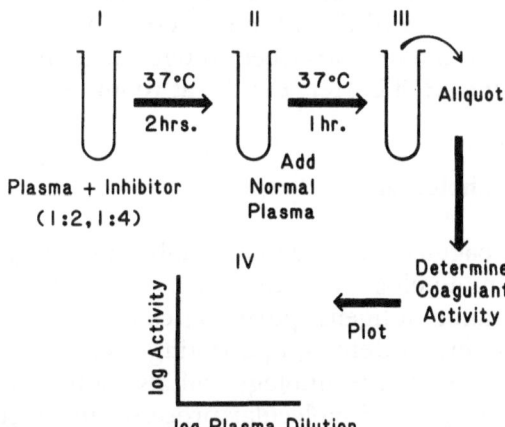

Figure 38. Diagramatic illustration of an antibody neutralization test.

inherited coagulation defects. These antibodies (inhibitors) neutralize the functional activity and are nonprecipitating. Using this technique and isoelectric focusing, molecular composition of inhibitors to Factors VIII and VII has been established. Hutlin et al and Koutts et al[48,58] have demonstrated that these inhibitors are composed of subpopulations of IgG. A general principle of the assay is outlined in Fig. 38.

Radioimmunoassay

Radioimmunoassay and immunoradiometric assays are used to quantitate low molecular weight peptide markers, such as fibrinopeptide A, $B\beta_{15-42}$, fragment E, platelet factor 4, 6-keto-$PGF_{1\alpha}$, $PGE_{2\alpha}$ and $PGF_{2\alpha}$ in investigating activation of the hemostatic pathways.

Immunoadsorption

Antibody-immobilized Sepharose gel columns are powerful tools in isolating clotting factors, as well as in removing contaminants in purified materials or therapeutic concentrates. Specific antibodies are adsorbed onto Sepharose gel by using a cross-linking agent, cyanogen bromide. At present, a wide variety of preactivated gel types are commercially available and can be employed in the procedure. Purified IgG, 100 mg, is mixed with activated Sepharose and stirred overnight at 4°C. Any unbound protein is washed with 0.2 M sodium carbonate buffer, pH 9.0, and 1 M ethanolamine, pH 9.0, is then added to block the unoccupied reactive sites in Sepharose. After 2 hours of incubation the mixture is washed with 0.5 M sodium carbonate buffer, pH 9.0, 0.1 M sodium acetate, pH 8.5, 0.1 M sodium borate, pH 4.1, containing 1 M NaCl, and finally by distilled water. The gel is now ready for immunoadsorption. The immunoadsorbed antigen is then dissociated by using 0.2 M glycine–HCl buffer, pH 2.8 at room temperature.

Conclusions

Coagulation proteins are highly immunogenic proteins; the sensitivity of immunologic reactions can be best utilized in quantitating low concentration antigens, purifying clotting factors, investigating the nature of inherited defects, and testing therapeutic concentrates. The advent of hybridoma technology will also contribute to the elucidation of many unexplained molecular processes in coagulation disorders.

Suggested readings

Hassouna HI, Penner JA: Antibody technics and blood coagulation. Semin Thromb Hemostasis 7(2):61–111, 1981.

Seegers WH: A personal perspective on hemostasis and thrombosis (1937–1981). Semin Thromb Hemostasis 7(3–4):178–307, 1981.

7

Quality Control

The primary objective of a quality control program in any clinical laboratory is to provide meaningful data on the health status of patients. This type of data can be utilized in the diagnosis and therapeutic management of a disease. In addition, an effective quality control procedure provides means of intralaboratory comparison of test results and forms a basis of epidemiologic investigation at a worldwide level. In order to acquire statistically significant data from an immunoassay, all sources of error due to procedure, variations in stability of reagents, and other assay parameters should be monitored by a rigid quality control scheme. Compared to other physicochemical analytic methods, immunologic techniques show a certain degree of imprecision. Batch-to-batch variations in antibody titer and a gradual degradation of antigen standard over a period of time present difficulties in the optimization of assay conditions. Nevertheless, performance characteristics of an immunoassay can be greatly improved by minimizing random error and reducing the systematic error by using the same reagents and assay parameters (temperature, incubation time, agarose medium, voltage, duration of electrophoresis and separation techniques). Each laboratory should recognize these variables and efforts directed toward eliminating sources of error. In our laboratories, we have optimized conditions for rocket immunoelectrophoresis, laser nephelometry, fluoroimmunoassay, and radioimmunoassay for quantitating, Factor VIII : RAg, enzyme-inhibitor complexes, antithrombin-III, α_2-antiplasmin, protein C, fibronectin, and other platelet specific proteins.

Immunologic Reagents

Standardization of immunologic reagents was adopted only during the last 10 to 15 years, when commercial antisera for IgG, IgM, and IgA became available. With the increasing use of antisera in quantitating proteins, it was soon realized that there was no uniformity of results between laboratories for the same sample, and a need for standard reagents was recognized. Standardization of reagents was considered at the level of purity, stability, and availability in large amounts. The choice of antibody standard was difficult because of their heterologous nature, and differences in avidity, specificity, and class type. Consequently antigen preparations which could meet the above criteria were adopted as reference materials.

Standard Preparation

The combined efforts of the International Union of Immunological Societies Standardization Committee, International Laboratory for Biological Standards, and several other national scientific agencies resulted in a set of requirements for the preparation of reference standards:

Table 20. Biological Characteristics of International Standards and Reference Preparations in Coagulation Testing

Standard	IU/Ampule	Form	Date Released	Designation
Ancord	55	In lactose and human serum albumin	1976	First reference preparation (RP)
Antithrombin-III	0.9	Freeze-dried in 1 ml normal plasma	1978	First RP
Blood coagulation Factor VIII, human	1.1	Freeze-dried concentrate	1976	Second standard
Blood coagulation Factor IX, human	5.62	Freeze-dried concentrate	1976	First standard
Heparin, porcine	1370	Freeze-dried	1973	Third standard
Plasmin	8.0	Solution in 50% glycerol	1976	First RP
Thrombin, human	100	Freeze-dried	1975	First RP
Thromboplastin, human	–	Freeze-dried	1976	First RP
Streptokinase	3100	Freeze-dried	1964	First RP
Urokinase	4800	Freeze-dried	1968	First RP

From WHO: Biological Substances, International Standards, Reference Preparations, and Reference Reagents. Geneva, 1979.

Table 21. Commercially Available Plasma Standards

Standard	Known Proteins	Source
ARP (Assayed Reference Plasma)	Broad spectrum	Helena
Coagent	Fibrinogen, Factors VIII and IX	Lancer
Kabi	Plasminogen, fibrinogen, AT-III	Kabi Group
NRP (Normal Reference Plasma)	Broad spectrum	George King
Omega	Broad spectrum	Hyland
Precichrom	AT-III, α_1-AT, α_2-M	Boehringer
Protein Standard Plasma	AT-III, plasminogen, fibrinogen, α_1-AT	Calbiochem
Quantitrol	α_1-AT, α_2-M	Kallestad
Thromboscreen	Broad spectrum	Pacific Hemostasis
Stago	Factors VIII : RAg and IX : RAg, fragments D and E	Stago

1. Possess the same biologic properties as the test sample to be quantitated.
2. Be homogeneous and free from contamination.
3. Be stable during long-term storage.
4. Should not denature during freeze-drying.
5. Reconstitute completely and give a clear solution.

A standard preparation which has these properties is essential to the success of a quality control program. After such a standard is prepared, it is submitted to the World Health Organization (WHO) for consideration by the Expert Committee on Biological Standardization. If the preparation meets the criteria set by the committee, a unit of activity is then assigned. The international unit (IU) for a specific protein is defined as the biologic activity contained in one ampule of freeze-dried material. Through this agency, reference preparations for Factor VIII, Factor IX, plasmin, thrombin, thromboplastin, antithrombin-III, urokinase, and streptokinase were introduced and are currently distributed by the National Institute for Biological Standards and Control in London. The biological characteristics of these reference preparations are shown in Table 20.

Presently several commercial plasma preparations have been cross-referenced with WHO or National Standards and contain a wide range of coagulation proteins with known concentration values (broad spectrum). These plasma standards, which are listed in Table 21, can be utilized as internal controls in evaluating performance characteristics of a given immunoassay.

Internal Standard

An in-house standard in our laboratories is prepared by drawing blood from ten male and ten female healthy volunteers into citrated tubes utilizing a double-syringe technique. Platelet-poor plasma is prepared by centrifuging the blood at 3000 rpm for 20 minutes. Clotting assays such as prothrombin time (PT), partial thromboplastin time (PTT), and thrombin time (TT) are performed on each plasma. Individual plasmas showing normal clotting time are pooled and stored in small aliquots at −70°C. This batch of pooled plasma is denoted as normal human pool (NHP), and can be used up to 3 months. However, as most coagulation proteins in NHP are stable only for 8–12 hours at 4°C, NHP should be used immediately after thawing.

Units of Concentration

For an internal control, NHP is assumed to contain 100% antigenic protein in all clotting factors, and serine protease inhibitors; patient samples are then quantitated in terms of percent NHP. This method of reporting results is adopted when rocket immunoelectrophoresis, laser nephelometry, fluoroimmunoassay, and enzyme-linked immunoassays are employed. Concentration of antithrombin-III, α_2-antiplasmin, α_1-antitrypsin, Factor VIII : RAg, prothrombin, and Factor XII in patient plasma is expressed in comparison to NHP. In radial immunodiffusion and RIA, concentration units are mg/dl, μg/ml, or ng/ml, depending upon the antigenic level present.

Procedure

The purpose of a quality control procedure is to measure and check on variations in an assay system. Once this variation is predicted, limits of variation and confidence intervals for the data are established. Statistical analyses in quality control were initiated in the early twentieth century: Shewhart and others[105,121] developed a technique of utilizing charts. This chart is known as the Shewhart chart and forms a basis of statistical control in experimental procedures. In addition to the Shewhart chart,

Youden and Cusum charts are also employed in monitoring quality assurance in immunology. A complete quality control scheme should include calibration of equipment, monitoring reagent purity and stability, and statistical evaluation of the data.

Calibration of Equipment

Calibration refers to accurate performance of equipment such as centrifuge, spectrophotometer, analyzers, and pH meters. A regular check on the operation of the following equipment will generally ensure good assay results.

1. *pH meter.* Since pH of the buffer plays an important role in antigen–antibody reactions, specific pH adjustments are required. The pH meter should be checked daily using a series of known buffers.
2. *Analytic balance.* The analytic balance should be placed on a solid, stable surface and calibrated with known standard weights certified by the National Bureau of Standards.
3. *Diluters.* All autodiluters and micropipettes should be examined for their accuracy at regular intervals.
4. *Spectrophotometers.* Check the proper wavelength and linearity in absorbence.
5. *Refrigerators and freezers.* A daily listing of temperature is recommended, as immunologic reagents may undergo degradation at unstable temperatures.
6. *Water baths.* All water baths should be equipped with a thermometer, and water temperature and level should be checked on a daily basis.
7. *Thermometers.* The accuracy of all laboratory thermometers should be verified against a precision certified thermometer.

Reagent Stability

To ensure stability all reagents should be stored at the proper temperature and thawed just before the assay. As most freeze-dried proteins require 10–15 minutes to go into solution after reconstitution, sufficient time should be allowed in the assay. All antisera preparations must be brought to room temperature before use for an effective antigen–antibody interaction. Stability of the standard should be evaluated over a period of storage by determining its concentration at three to four levels using a monospecific antiserum. In-house controls are treated similarly and the values are recorded on a Shewhart chart.

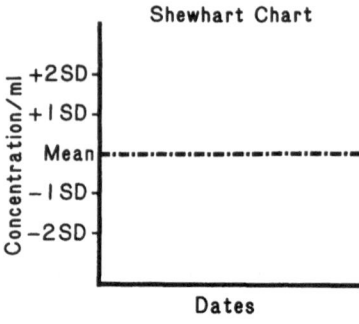

Figure 39. A typical Shewhart chart used in a quality control program.

Quality Control Charts

Shewhart Chart

The Shewhart chart is maintained in order to establish internal control values with reference to an international standard. Protein concentration in a control sample is measured over 20–30 days, mean and standard deviation are established, and the chart (Fig. 39) is set up with the mean and one to two standard deviation levels on both sides of the mean. If the performance of the assay is satisfactory, control values will be evenly distributed around the mean; deterioration in control or standard is indicated otherwise.

Youden Chart

The Youden chart is similar to the Shewhart chart with the exception that the former requires determination of control values at two levels of

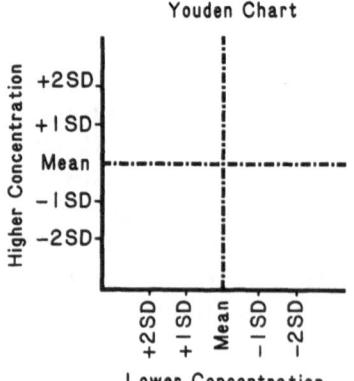

Figure 40. Youden chart which monitors the lower and higher concentration of a protein around the two means.

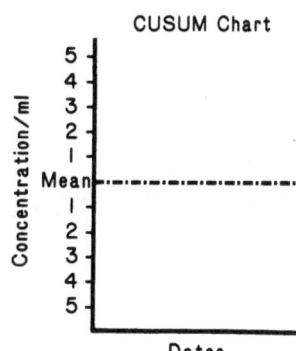

Figure 41. The cumulative sum chart is used to monitor the accuracy of the assay.

concentration; these are plotted as mean values against one to two standard deviations. Efficiency of the assay is indicated by an even distribution of values about the center (Fig. 40).

Cusum Chart

The cumulative sum (Cusum) chart is designed to test the accuracy of the assay. As in other monitoring systems, the in-house control or international standard is measured on several occasions and a mean is calculated. The difference between individual determination value and the mean is plotted (Fig. 41).

After any one of these quality control charts is maintained in a laboratory, data are further analyzed for percent coefficient of variation. For methods in statistical analysis, the reader is referred to Westgard et al.[118]

Suggested readings

Inhorn SL (ed): Quality Assurance Practices for Health Laboratories. Washington, D.C., American Public Health Association, 1978, pp 745–787.

WHO: The Collection, Fractionation, Quality Control, and Use of Blood and Blood Products. Geneva, 1981.

8

Future Outlook

Laboratory testing of coagulation disorders has come a long way from bleeding time, clotting time, prothrombin time, and partial thromboplastin time tests to the detection of nanogram and picogram/milliliter amounts of fibrinopeptide A, beta-thromboglobulin, platelet factor 4, and $B\beta_{15-42}$ peptide. Earlier diagnoses were impeded by a lack of understanding of the biochemistry of blood coagulation and the nature of coagulation enzymes. In 1905, Morawitz described his "classic theory" of blood coagulation in which he proposed a two-step mechanism: (1) prothrombin to thrombin and (2) fibrinogen to fibrin conversion. It was only after the discovery of additional clotting factors, coumarin anticoagulants, heparin cofactor, heparin therapy, and the role of vitamin K in clotting factor synthesis that a new picture of the coagulation mechanism began to emerge.

Several biochemical and immunochemical assays were later developed to study the pathophysiology of hemostatic functions. Until recently, coagulation disorders were mainly studied by measuring absolute concentration of a protein and its biologic activity. Functional activity of clotting factors and inhibitors was considered the most important laboratory parameter and diagnosis was based largely on such determinations. Quantitative functional assays for coagulation and fibrinolytic enzymes depended on determining rate of formation or dissolution of a fibrin clot. Clotting assays such as prothrombin time, activated partial thromboplastin time, and thrombin time are still utilized extensively in the diagnosis and management of various forms of coagulopathies. For example, oral anticoagulant therapy, heparin therapy, thrombolytic therapy, diagnosis of disseminated intravascular coagulation, hypercoagulable states, liver diseases, hemophilia A and B, and circulating anticoagulants are all currently monitored by use of these clotting tests.

Immunodiagnosis of coagulation disorders is made by employing radial immunodiffusion, double diffusion, immunoelectrophoresis, and rocket immunoelectrophoresis. It is now increasingly recognized that functional activity of a plasma protein is not always comparable to its immunologic level. This dichotomy has to be considered when evaluating a particular coagulation defect. Molecular abnormalities in fibrinogen, prothrombin, Factor VIII, Factor IX, and in various vitamin K–dependent factors have been established on the basis of differences in functional vs. immunologic properties. With this approach, a new investigative dimension will be added to the spectrum of methods in coagulation testing. With improved methods of protein purification and monoclonal antibody production techniques, highly purified antigens and specific antisera to a large number of coagulation proteins have become available. Nonisotopic immunoassays such as fluoroimmunoassay, enzyme-linked immunoassay, and rate and laser nephelometric assay have been developed and are utilized to measure antithrombin-III, α_2-antiplasmin, plasminogen, Factor VIII : CAg, Factor VIII : RAg, Factor IX, and fibrinogen degradation products (FDP).

In addition, radioimmunoassays have contributed a great deal to the present understanding of the mechanisms of thrombosis and its diagnosis. For example, thrombin cleaves fibrinogen into two molecules of fibrinopeptide A (FPA), two molecules of fibrinopeptide B (FPB), and fibrin monomers. Soon after these fibrinopeptides are released, fibrin monomers polymerize to form a fibrin clot. Laboratory diagnosis of thromboembolic diseases is usually performed by contrast venography, impedence plethysmography, phleborrheology, Doppler technique, and ^{125}I-fibrinogen leg scanning. Besides being expensive and painful to the patient, these tests only indicate the anatomic extent of thrombus formation. A much simpler assay is to measure FPA in the blood, which would indicate the rate of conversion of fibrinogen to fibrin and might be useful in the diagnosis of deep vein thrombosis and other thrombotic disorders. Since FPA is a smaller molecular weight peptide (MW 1535 daltons), antibodies raised against these peptides do not precipitate in gels and hence cannot be used in immunodiffusion techniques.

RIA is the only immunotechnique that can detect antigenically active structures in fibrinogen. Thromboxane B_2, the metabolic derivative of thromboxane A_2, a potent prostaglandin found in platelets during activation, can be measured in serum by RIA as can 6-keto-PFF$_{1\alpha}$, a derivative of the platelet inhibitor prostacyclin. Factor VIII : CAg and Factor VIII : RAg (low in von Willebrand's disease) are now quantitated by RIA techniques in order to distinguish between hemophilia and von Willebrand's disease. Fibrinopeptide A and fragment E antigen assays have been developed to evaluate hypercoagulable, disseminated intravascular coagulation and related fibrinolytic processes.

In the past, automation of immunologic methods in coagulation testing was not attempted primarily due to lack of appropriate instrumentation. However, with the recent introduction of kinetic laser nephelome-

Figure 42. A fully automated laser nephelometric system capable of quantitating coagulation proteins. Courtesy of Calbiochem Behring.

ters (Fig. 42), centrifugal analyzers, and advanced spectrophotometers, automation of immunologic techniques will soon become a reality. Fluoro-, enzyme-linked immuno-, and radioimmunoassays will be performed on such automated systems as Gilford PR-50, Quantum II, IL-Multistat-III, RA-1000, and so on for the quantitation of Factor VIII : RAg, Factor IX, fibrinopeptide A, $B\beta_{15-42}$, and platelet specific proteins, which will aid in the assessment and therapeutic management of hemorrhagic and thrombotic diseases on an immediate basis.

Conclusions

With the emergence of newer immunochemical technologies automation in coagulation testing has become a strong possibility. Many clinical chemistry analyzers and computer-assisted instruments can be easily adapted to perform nephelometric, fluorometric, enzyme-linked, and radioimmunoassays for almost all coagulation factors and serine protease inhibitors. In addition to short turn-around time, such procedures will ensure good quality control and standardization practices, which are greatly lacking in this area of hemostatic testing. Automation will also play a key role in the development of newer coagulation profiling tests, monitoring of anticoagulant therapy, and the diagnosis of cardiovascular diseases.

Suggested reading

Bick RL: Clinical hemostasis practice: The major impact of laboratory automation. Semin Thromb Hemostasis 9(3):139–171, 1983.

Appendix A

Techniques in Antibody Production

The immunogenic properties of an antigen largely depend on its molecular weight, configuration (globular vs. fibrillar), and rates of catabolism and elimination by the host animal. Also, the antigenic determinant sites increase as the molecular weight of the antigen increases. Proteins, glycoproteins, and nucleoproteins are effective immunogens, whereas lipids, nucleic acids, and carbohydrates are weakly antigenic. Since most of the coagulation factors and serine protease inhibitors are high molecular weight proteins/glycoproteins, they are capable of mounting an effective antibody response when injected into animals. Also, since many of these proteins can be isolated and purified on various affinity chromatographic columns, monospecific antibodies can be raised against them. Such animals as rabbits, monkeys, goats, pigs, and guinea pigs may serve as host animals. In general, New Zealand white male rabbits are quite suitable for producing specific antisera to antithrombin-III, α_1-antitrypsin, α_2-antiplasmin, protein C, and other coagulation factors. By a simple process of ammonium sulfate precipitation and subsequent gel filtration on an ion exchange diethylaminoethyl (DEAE) Sepharose column, IgG fractions of antibodies can be obtained. This section will briefly discuss techniques of isolation and purification of some coagulation proteins from plasma and raising of antiserum in rabbits.

Antigen Purification

Isolation and purification of blood clotting factors and inhibitors is performed by using affinity chromatographic techniques. Most of the vitamin K–dependent factors (II, VII, IX, and X) can be adsorbed on barium salts, aluminum hydroxide, and dextran sulfate. Antithrombin-III (AT-III) binds to heparin when plasma is filtered through a heparin–Sepharose column, and is eluted from the Sepharose column by a salt gradient. The other inhibitor, α_1-antitrypsin (α_1-AT), is purified on a Concanavalin-A Sepharose affinity column and is eluted with a sodium acetate buffer containing 50 mM α-methyl-D(+)-glucoside. Plasmin inhibitor α_2-antiplasmin (α_2-AP) can be isolated by immunoadsorption chromatography on 6-aminohexyl-Sepharose substituted with anti–α_2-antiplasmin and further purified by Ultragel AcA 44 filtration. Several affinity gel preparations are now available from Bio-Rad, Pharmacia, LKB, and other sources. A simple procedure for the isolation of AT-III, α_1-AT, and α_2-AP as adopted in our laboratory is given below:

Isolation of AT-III

1. Two grams of heparin–Sepharose CL-6B (Pharmacia) is swollen in 0.05 M Tris–HCl buffer, pH 7.4, overnight.
2. Fill a 20-cm Econo-column (Bio-Rad) with the swollen gel to a height of 12 cm.
3. Equilibrate the column by repeated washings of the gel with buffer.
4. Load the column with 1 ml of freshly prepared normal human plasma and let it go through the gel.
5. Wash the column with 0.05 M Tris–HCl buffer and collect 15 0.5-ml fractions. These can be discarded.
6. Elute the adsorbed AT-III with a 2 M NaCl solution and obtain 15 0.5-ml fractions.
7. Determine OD_{280} of the fractions, plot OD_{280} vs. fraction numbers on linear graph paper, and pool fractions showing AT-III activity.
8. To determine AT-III concentration, commercial radial immunodiffusion plates for AT-III can be used. AT-III activity can be measured by a synthetic chromogenic substrate method.
9. Desalting of fractions is performed by using P-10 Pharmacia columns. Purity of this preparation can be tested in an immunoelectrophoresis system utilizing a monospecific anti–AT-III.
10. The preparation can be further purified on an ion exchange column, depending upon the nature of the investigation.

Isolation of α_1-AT

1. Fill a 20-cm Econo-column (Bio-Rad) with a Concanavalin-A Sepharose 4B (Pharmacia) preswollen slurry.
2. Equilibrate the column with 0.05 M sodium acetate buffer, pH 7.0, with repeated washings.
3. Load the column with 1 ml of freshly prepared normal human plasma and let it pass into the gel.
4. Wash the column with the buffer and collect 15 0.5-ml fractions. These can be discarded.

5. Elute the adsorbed α_1-AT with the washing buffer containing 50 mM α-methyl-D(+)-glucoside, and collect 15 0.5-ml fractions.
6. Determine OD_{280} of the fractions and plot OD_{280} vs. fraction numbers.
7. Determine α_1-AT concentration by using a commercial RID plate for α_1-AT. Pool fractions showing positive results in RID plate.
8. Desalt the fractions by using P-10 Pharmacia columns.
9. Test purity of the material in an immunoelectrophoresis system using a monospecific anti–α_1-AT.

Isolation of Plasminogen

1. Dissolve 10 g L-lysine monohydrochloride (Sigma) in 25 ml of 0.1 M $NaHCO_3$, and adjust the pH to 8.9.
2. Weigh 10 g freeze-dried CNBr-activated Sepharose 4B (Pharmacia) and allow it to swell in 100 ml of 1 mM HCl. Wash the gel repeatedly with 1 mM HCl and finally resuspend it in 40 ml of 1 mM HCl.
3. Add 25 ml L-lysine monohydrochloride to 40 ml of CNBr-activated Sepharose 4B and stir the solution overnight at 4°C.
4. Remove the uncoupled lysine by repeated washings with 1 M NaCl and then with 0.1 M $NaHCO_3$.
5. Store the lysine–Sepharose slurry at 4°C.

Chromatography

1. Fill a 2 × 30 cm column with 35 ml lysine–Sepharose 4B slurry and equilibrate the column with 0.1 M sodium phosphate–2 mM EDTA buffer, pH 7.4.
2. Pass 25 ml of outdated Blood Bank normal plasma through the column and wash the column with buffer until the absorbency at 280 nm is less than 0.05.
3. Elute plasminogen with 0.1 M lysine and remove lysine by extended dialysis against distilled water.
4. Store this plasminogen preparation at −70°C.

Isolation of α_2-Antiplasmin

Plasminogen-depleted plasma can be further utilized to isolate α_2-antiplasmin.*

---

* The procedure is modified from Moroi et al: J Biol Chem 251:5956, 1976.

1. Add solid ammonium sulfate to plasminogen-depleted plasma to a final concentration of about 40% saturation.
2. Centrifuge the solution at 3000 rpm for 20 minutes and dissolve the precipitate in 0.1 M sodium phosphate buffer, pH 7.4.
3. The same procedure can be repeated with the supernatant.
4. Dialyze this solution against the phosphate buffer overnight at 4°C.

Chromatography

1. Fill a small (2 × 14 cm) column with an ion exchange medium (DEAE–Sephadex A-50 [Pharmacia]) and equilibrate the column with 0.05 M phosphate buffer, pH 7.4.
2. Add the dialyzed fraction on top of the column and allow it to pass through the gel.
3. Wash the column with buffer and elute α_2-antiplasmin with 2 M NaCl.
4. Fractions showing inhibitor activity (determined by a synthetic chromogenic substrate method) are pooled, dialyzed, and stored at $-70°C$.
5. Further purification can be achieved by using plasminogen–Sepharose affinity chromatography.

Protein Determination

Protein determination is an important aspect of laboratory investigations and, depending upon the need, a number of methods are available. In a coagulation laboratory, total protein concentration in standard solutions, serum, plasma, and cellular extracts may be conveniently determined by any of these methods.

Micro-Kjeldahl Method

Protein is broken down to ammonia and the resulting ammonia concentration is determined by the phenol–hypochlorite reaction.

Reagents

1. Alkaline hypochlorite. Dissolve 12.5 g sodium hydroxide in 400 ml distilled water; cool, add 20 ml sodium hypochlorite (any bleach solution), and make it to 500 ml. Store in a polyethylene bottle at 4°C.

2. Phenol solution. Dissolve 25 g phenol and 0.05 g nitroprusside in 500 ml distilled water. Store in a brown bottle at 4°C.
3. Extraction mixture. Carefully add 30 ml sulfuric acid to 50 ml distilled water and cool. Add 30 g anhydrous sodium sulfate and make it to 100 ml. Add 0.5 g mercuric chloride and mix well.
4. Sodium hydroxide. Dissolve 100 g sodium hydroxide in water, cool, and make it to 1 liter.
5. Ammonia standard. Dissolve 0.47 g ammonium sulfate in water and add a few drops of concentrated sulfuric acid. Dilute it to 1 liter. This solution contains 0.1 mg nitrogen/ml. Dilute it 1 : 5 to obtain 20 μg/ml concentration.

Procedure

1. Make a 1 : 100 dilution of the sample with saline.
2. Add 1 ml of this sample to a graduated tube and mix it with 1 ml of the extraction mixture.
3. Add 2 small glass beads, cover the tube lightly with a glass stopper, and heat the mixture for about 30 minutes.
4. Mix 1 ml of saline and 1 ml of extraction mixture to be used as a blank, in a separate tube. Heat the mixture as above.
5. Add 10 ml water, mix, and add 4 ml of sodium hydroxide to sample and saline control tubes. Cool the solution and add enough water to make 50 ml.
6. In a separate tube, make a 1 : 10 dilution of the 20 μg/ml standard.
7. Arrange tubes for sample, standard, and blank. Pipette 1 ml of extracted sample, standard, and blank into respective tubes.
8. Add 1 ml of phenol solution to these tubes, mix well, and add 1 ml alkaline hypochlorite. Mix the tubes well and incubate at 37°C for 20 minutes.
9. Add 10 ml distilled water to each tube, mix the contents, and read OD_{625} in a spectrophotometer, against the blank.
10. Calculation:

$$\frac{\text{Absorbence of sample}}{\text{Absorbence of standard}} \times 1 = \text{g protein nitrogen}/100 \text{ ml}$$

Refractometer Method

In most laboratories, rapid estimation of total proteins is made by using refractometers. American Optical Corporation's TS meter is a hand refractometer which reads the concentration values in grams protein per 100 ml.

Dye Binding Method

Although protein measurement with Lowry's method* is most commonly utilized, a relatively convenient dye binding method has been introduced recently and is commercially available from Bio-Rad Laboratories. The assay originally described by Bradford† utilizes color change of a dye in response to various concentrations of protein. The dye is available from Bio-Rad. The procedure is summarized below:

1. Bovine serum albumin: standard A (1 mg protein/ml)
                         standard B (10 mg protein/ml)
2. Standard curve:

μg/tube	μl A	μl B	Dye (5 ml)	OD₅₉₅
10	10			
20	20		+	
40	40		+	
60	60		+	
80	80		+	
100	100		+	
200	200	20	+	
300		30	+	
400		40	+	
500		50	+	

3. Unknown sample: mix 100 μl with 5 ml dye.
4. Control: mix 100 μl saline with 5 ml dye and read OD₅₉₅ against this blank. Determine protein concentration with reference to the standard curve.

Immunization

Antigen Preparation

In order to enhance the immunogenic properties of a given antigen, unrelated nonantigenic substances are injected along with the antigen. The extent and duration of immune response is then greatly increased because of these substances. Such agents are referred to as adjuvants. Freund's complete adjuvant, which is a mixture of killed *Mycobacterium tuberculosis,* mineral oil, and a detergent is the most widely utilized adjuvant.

---

* See J Biol Chem 193:265–275, 1951.
† See Anal Biochem 72:242–254, 1976.

Preparation of Freund's Adjuvant

Freund's adjuvant can be used as a complete adjuvant or an incomplete adjuvant depending upon the presence or absence of the bacteria.

Incomplete adjuvant

The incomplete adjuvant can be prepared as follows:

1. 3 parts light mineral oil, U.S.P. standard.
2. 1 part Aquaphore, Falba, or anhydrous lanolin.
3. 4 parts phosphate buffer.

Complete adjuvant

The complete adjuvant contains killed bacterial preparation *Mycobacterium tuberculosis*, 10 mg/ml, in the above oil emulsion.

For immunization of an animal, equal volumes of antigen and Freund's complete adjuvant are emulsified using a glass syringe and a beaker; the degree of emulsification is tested by observing the disintegration of a drop of this emulsion in a layer of water. If the drop stays intact, the preparation is ready for immunization. The amount of antigen to be injected depends upon the antigenicity, purity, and molecular weight of the protein. Initially 20–30 μg of purified AT-III, 200–300 μg α_1-AT, or 100–200 μg Factor VII can be injected intramuscularly or intradermally into foot pads (Fig. 43) at weekly intervals for 2–3 weeks, followed by a booster dose. Blood samples can be analyzed for antibody production 1 week after the booster. Hyperimmune serum may be prepared by continuing the injections every month and bleeding the animal once every 15 days. However, these guidelines (for amounts of antigen to be used) may differ with the antisera purification procedures adopted in one's laboratory and the route of injection. A total of 1–10 mg protein can be safely administered.

Animals

New Zealand white male rabbits weighing 2–3 kg body weight are suitable for raising antisera to most coagulation proteins. Rhesus monkeys may also be used for this purpose, but rabbits are easier to handle, besides being cheaper than monkeys. It is advisable to immunize 3 or 4 animals with one antigen at the same time in order to get a large supply of high titer antibodies.

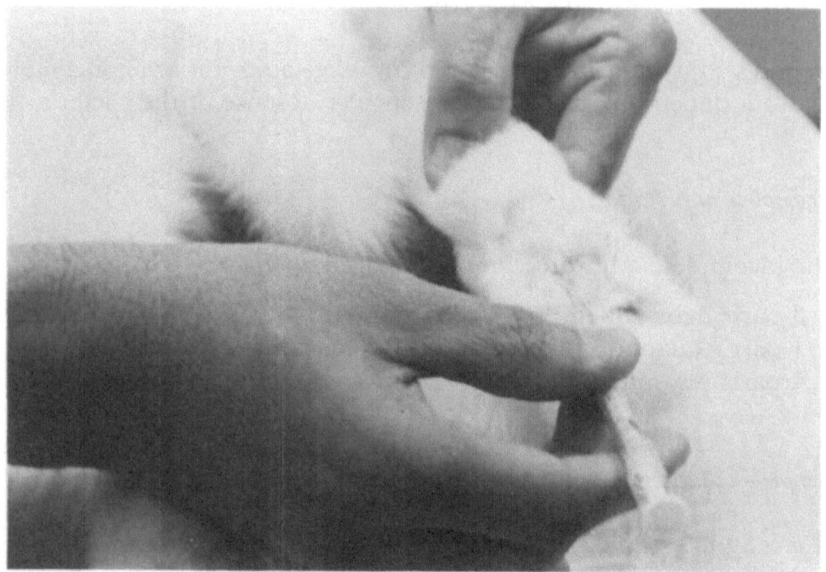

Figure 43. Antigen administration into foot pad of New Zealand white male rabbits.

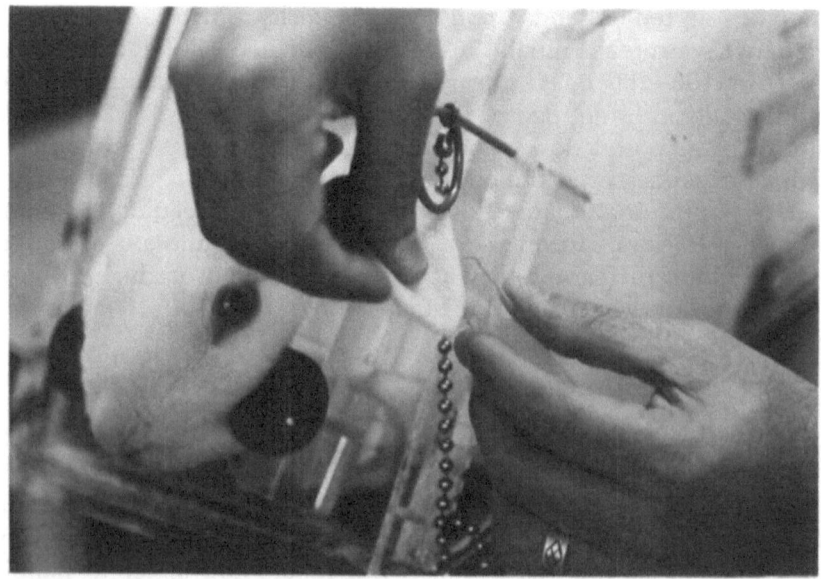

Figure 44. Collection of blood by making an incision into the marginal ear vein. With this technique 10–20 ml of blood can be obtained at a time. The rabbit is secured in the box.

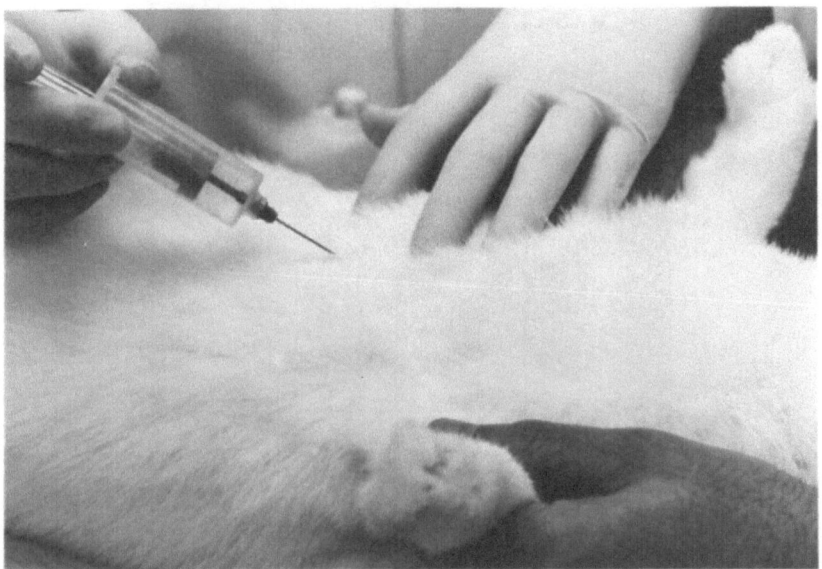

Figure 45. A large amount of blood can be obtained by cardiac puncture, before the animal is sacrificed.

Bleeding

Bleeding animals to obtain immune serum is effected by various methods. The most common method is the ear vein puncture, in which the ear is first irritated by rubbing xylene on it and making an incision in the contralateral vein. A sterile test tube is then placed close to the ear and blood is allowed to drip directly into the tube (Fig. 44). After the end of the procedure, a slight pressure is applied to stop the flow. Although the procedure is simple, blood drips slowly and 5–10 ml of blood can be obtained for primary testing. The incision is often clogged. Cardiac puncture (Fig. 45) is a rapid method, but requires precision. The puncture is performed by forcing a 1-mm cannula on a syringe through the frontal and left side of the thorax between the fourth and fifth ribs. Blood is drawn when the needle reaches the heart cavity. Thirty to 40 ml of blood can be obtained from a large rabbit for antiserum. Blood is collected without any anticoagulant.

Preparation of Antiserum

After the blood is collected, it is allowed to clot for 30–60 minutes at room temperature and then centrifuged at 3000 rpm for 15 minutes in a refrigerated centrifuge. The clear serum is collected into sterile glass

Figure 46. A dialysis chamber showing the cellulose membrane containing ammonium sulfate-fractionated antiserum. Ammonium sulfate, salts, and small molecular weight proteins will diffuse through the cellulose membrane into the dialysis buffer.

tubes, and an equal amount of 40% saturated ammonium sulfate is added. The mixture is refrigerated overnight for complete precipitation of immunoglobulins, centrifuged at 3000 rpm for 15 minutes, and the supernatant, which contains albumin, haptoglobulin, transferin, hemoglobin, and other α-proteins, is discarded. The precipitate is dissolved in a 0.05 M phosphate buffer, pH 8.0. Ammonium sulfate and other impurities are removed by a process of cellulose membrane dialysis (Fig. 46) against deionized distilled water and finally against 0.05 M phosphate buffer, pH 8.0.

Purification on Protein A–Sepharose Column

Protein A–Sepharose CL-4B is utilized to isolate the IgG-type antibodies. Protein A is isolated from cultures of *Staphylococcus aureus* and is known to interact with the Fc portion of IgG molecules from various

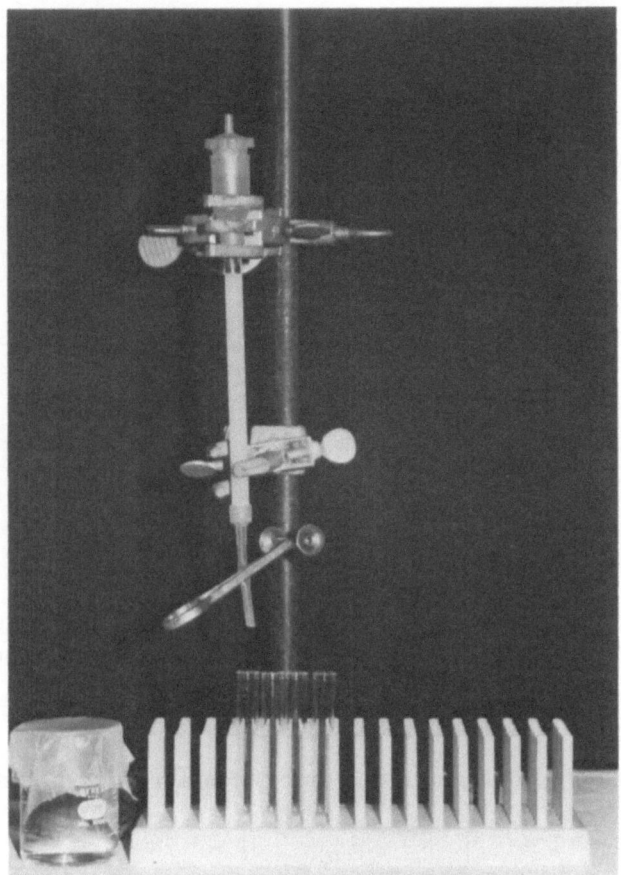

Figure 47. Isolation of IgG-type antibodies on protein A—Sepharose column.

species. The IgG-type antibodies can be isolated by passing dialyzed antiserum onto a protein A—Sepharose column and eluting it with 1 M acetic acid (Fig. 47). The antibodies are then dialyzed against deionized distilled water and the pH is adjusted to 8.0 by using a strong buffer.

Testing of Antiserum Activity

Microscope slides are coated with 2–3 ml of 1% agarose and gel is allowed to cool. Using Gelman gel cutters, a set of wells is cut and 5 μl of diluted antibodies is incorporated into the central well. An equal amount of antigen is incorporated into the outer wells and the glass slide is incubated in a moist chamber at room temperature.

Precipitin areas are visible after 12–14 hours of incubation. In a

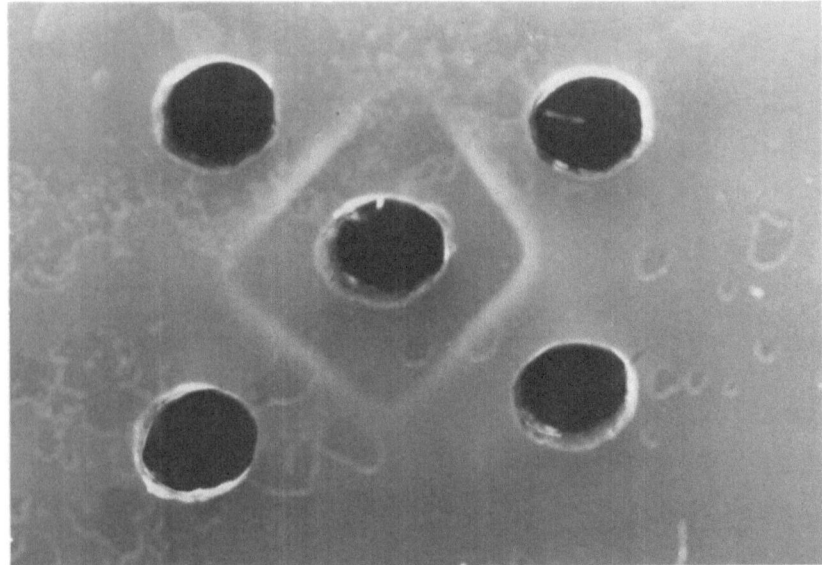

Figure 48. A double diffusion technique for testing purified antiserum. The central well contains purified anti–AT-III and the outer wells contain normal human pooled plasma (NHP). Only one precipitin arc is visible.

highly purified preparation, only one precipitin band should be visible (Fig. 48).

FITC-labeling of Antibodies

Purified antibodies can be conjugated to fluorescent dyes such as fluorescein isothiocyanate (FITC) by various techniques. A simple procedure consists of dissolving 100 μg of FITC dye in 2 ml of 0.1 M Na_2HPO_4 buffer, pH 9.2, by constant stirring with a magnetic bar. In another serum vial, 1 ml of 0.2 M Na_2HPO_4 is mixed with 1 ml of purified antibody (5 mg) and the FITC dye is then added. The total volume is adjusted to 8 ml by using normal saline. The mixing of solution is continued for 18 hours at 4°C, the solution dialyzed for 18 hours against 0.15 M NaCl, and the pH is adjusted to 7.4 with 1 N HCl. The labeled antibody is separated by gel filtration on a Sephadex G-200 column using phosphate buffered saline. All fractions showing orange color are collected and stored in small aliquots at −70°C.

Appendix B

Blood Collecting Techniques

Proper blood collecting technique is an important requisite to a reliable quantitation of plasma proteins and assessment of coagulation disorders. Depending upon the nature of proteins to be quantitated in an immunoassay, blood should be drawn in tubes containing appropriate anticoagulants. Sodium citrate in a 9 : 1 ratio of blood to anticoagulant preserves labile coagulation factors and is utilized for preparing plasma for factor assays and other serine protease inhibitors. However, to quantitate platelet specific proteins, i.e., platelet factor 4 and beta-thromboglobulin, ethylenediamine tetraacetic acid (EDTA) is used as an anticoagulant. Special agents such as indomethacin or transylol are required for $B\beta_{15-42}$ and TxB_2. In addition to the proper use of anticoagulants, a precise draw of the blood in a 9 : 1 ratio is essential; a short draw will result in abnormal values.

A brief outline of the blood collecting procedure is presented below.

Double Syringe Method (Fig. 49)

1. Select the arm to be used for drawing blood from a vein.
2. Tighten the tourniquet on the arm to make the vein prominent.
3. Cleanse the area with 70% isopropyl alcohol and let it dry.
4. Using a 5-ml plastic syringe and a 19- or 20-gauge butterfly infusion set (Abbott), enter the vein with minimum trauma.
5. Draw 2–3 ml of blood and change the syringe. Discard this draw of blood since it contains tissue thromboplastins.
6. With the second syringe draw 9 ml of blood.
7. Release the tourniquet, withdraw the needle, place a sterile gauze on the punctured area, and apply gentle pressure.
8. Withdraw the needle from the syringe, expel blood into the appropriate tubes containing 1 ml anticoagulant, and mix slowly.

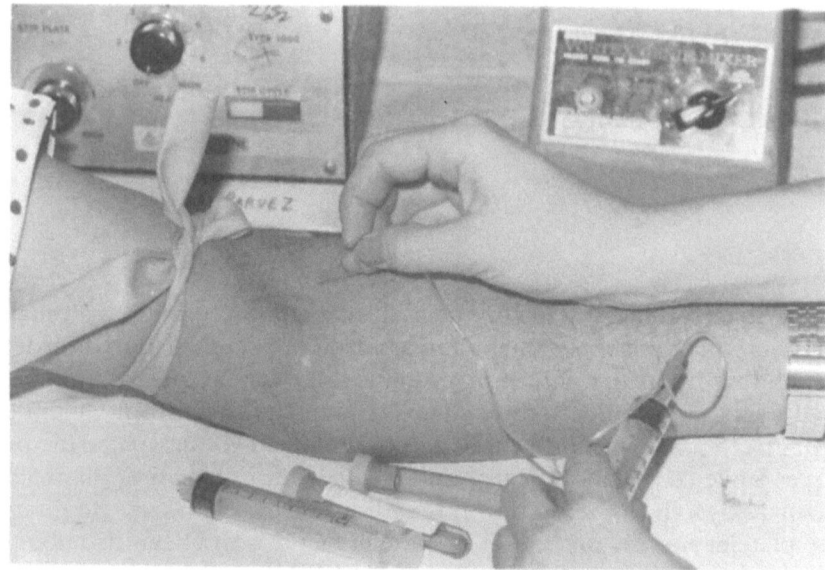

Figure 49. Blood collecting technique showing utilization of a Butterfly infusion set and two plastic syringes. The first draw of 2–3 ml is discarded and another 10 ml is drawn by the second syringe.

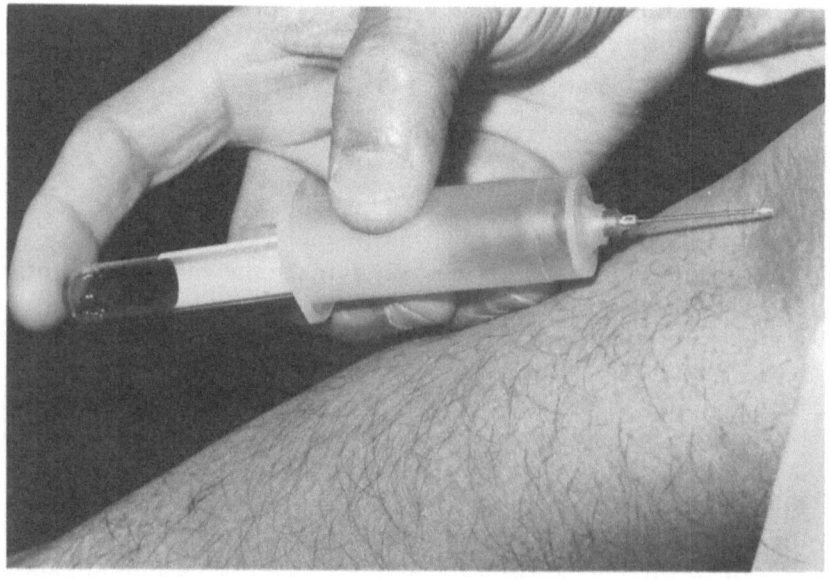

Figure 50. Blood can also be collected by a vacutainer method in which evacuated tubes and a needle with a shutoff valve are used, instead of the Butterfly infusion set.

Plasma Preparation

1. Centrifuge anticoagulated blood at 3000 rpm for 15 minutes in a refrigerated centrifuge.
2. Collect clear supernatant with a plastic pipette and store in small aliquots at −70°C or use immediately.

Vacutainer Method (Fig. 50)

Blood can also be drawn by using a vacutainer technique in which an evacuated tube is placed in a plastic holder. A suitable needle is then attached to the holder. The needle is inserted into the vein and the evacuated tube is gently pushed inside the holder to break the vacuum in the tube. The blood starts flowing into the collection tube. Commercially available blood collecting tubes are color coded to indicate the type of anticoagulant. A blue top, for example, indicates sodium citrate; purple top, EDTA; green top, heparin; red top, no anticoagulant.

Appendix C

Buffers and Stains

Buffers and stains are important constituents of an immunoassay and should be carefully selected and prepared. A guide for the preparation and use of buffers in biologic systems is available from Calbiochem Behring Corporation.[42]

Phosphate Buffer

In order to prepare 0.1 M phosphate buffer with different pH requirements, stock solutions of monobasic (A) and dibasic (B) sodium phosphate are prepared. By mixing precalculated amounts of A and B stock, solutions of the required pH can be prepared. Table 22 shows the amounts of A and B to be mixed. Stock solution A is made by dissolving 27.6 g monobasic sodium phosphate monohydrate in 1000 ml deionized water. Stock solution B is prepared by dissolving 28.4 g dibasic sodium phosphate in 1000 ml of deionized water.

Veronal, Barbital, or Barbitone Buffer

Barbitone buffer, 0.05 M, pH 8.4 is commonly used in electrophoresis. It is prepared by dissolving 42.5 g sodium diethylbarbiturate (sodium veronal) in 58.0 ml 1 N HCl and the total volume is adjusted to 5.0 liters with distilled water.

Veronal Acetate Buffer, pH 5.6–8.3

The veronal acetate buffer can be prepared from the following three stock solutions:

Table 22. Preparation of a 0.1 M Phosphate Buffer by Mixing Monobasic and Dibasic Sodium Phosphate Solutions

Monobasic (ml)	Dibasic (ml)	pH
92.0	8.0	5.8
90.0	10.0	5.9
87.7	12.3	6.0
85.0	15.0	6.1
81.5	19.5	6.2
77.5	22.5	6.3
73.5	26.5	6.4
68.5	31.5	6.5
62.5	37.5	6.6
56.5	43.5	6.7
51.0	49.0	6.8
45.0	55.0	6.9
39.0	61.0	7.0
33.0	67.0	7.1
28.0	72.0	7.2
23.0	77.0	7.3
19.0	81.0	7.4
16.0	84.0	7.5
13.0	87.0	7.6

1. Sodium acetate, anhydrous, 6.276 g
   Sodium diethyl barbiturate, 15.77 g
   1.0 M $MgCl_2$, 0.5 ml
   0.3 M $CaCl_2$, 0.5 ml
   Distilled water, 11 ml
2. 1 N HCL, 38.30 ml
   1 M $MgCl_2$, 0.25 ml
   0.3 M $CaCl_2$, 0.125 ml
   Distilled water, 250 ml
3. NaCl, 9.01 g
   1 M $MgCl_2$, 0.5 ml
   0.3 M $CaCl_2$, 0.5 ml
   Distilled water, 11 ml

To prepare a 0.15 M veronal acetate buffer, the three stock solutions are mixed in the following proportion:

Solution B (ml)	Solution A (ml)	Solution C (ml)	pH	M
5.15	10	84.85	5.6	0.148
4.92	10	85.08	6.13	0.148
4.65	10	85.35	6.70	0.148
4.20	10	85.80	7.08	0.149
3.0	10	87.0	7.65	0.151
0.20	10	89.80	8.38	0.155

Phosphate Buffered Saline (PBS)

PBS is prepared from the following reagents:

Solution A: $NaH_2PO_4 \cdot H_2O$, 27.6 g/liter
Solution B: $Na_2HPO_4 \cdot 12\ H_2O$, 71.63 g/liter
NaCl, 8.5 g/liter

To prepare 0.05 M PBS, pH 7.1, mix 70 ml solution A, 180 ml solution B, 5.7 g NaCl, and distilled water to 1 liter total volume.

0.05 M Tris Buffer, pH 7.2–8.6

Make a 50-ml stock solution of Tris–HCl (A) by dissolving 1.58 g of Tris–HCl in distilled water and prepare a 50-ml stock solution of Tris–base (B) by dissolving 1.21 g Tris–Base in distilled water. Mix the two solutions in the correct proportions to obtain 0.05 M buffer of desired pH (Table 23). After mixing the solutions, add 200 ml distilled water.

Table 23. Preparation of a 0.05 M Tris Buffer by Mixing Tris–HCl (A) and Tris–Base (B) Solutions

Stock A (ml)	Stock B (ml)	pH (ml)
44.2	50	7.2
41.4	50	7.4
38.4	50	7.6
32.5	50	7.8
26.8	50	8.0
21.9	50	8.2
16.5	50	8.4
12.2	50	8.6

Stains

Coomassie Blue

Coomassie Blue, 0.05 g
Glacial acetic acid, 9.2 ml
Methyl alcohol, 45 ml
Distilled water to 100 ml

Dissolve Coomassie Blue in 9.2 ml glacial acetic acid, and then add methyl alcohol. The stain should dissolve completely; otherwise use a magnetic stirrer. Add distilled water to make 100 ml. The stain can be used repeatedly.

Amido Black

Amido black 10 B, 0.5 g
Mercuric chloride, 5.0 g
Glacial acetic acid, 5.0 ml
Distilled water to 100 ml

Filter the stain before use. Destaining is achieved by washing with 2% acetic acid.

Azocarmine B (for cellulose acetate)

Azocarmine B, 0.05 g
2% (v/v) glacial acetic acid in distilled water, 100 ml

Stain for 1 hour. Destain in 2% acetic acid.

Nigrosin (for cellulose acetate)

Nigrosin, 5 mg
2% glacial acetic acid in distilled water, 100 ml

Stain for 1 hour. Destain with 2% acetic acid. Dry the membrane between filter paper.

Sudan Black B (for lipids)

1. Make a saturated solution of Sudan Black B in 60% ethanol at 37°C.
2. Cool the solution and filter.

3. Add 0.1 ml 25% sodium hydroxide to 50 ml of the stain.
4. Stain the protein for 1–2 hours.
5. Destain in 50% ethanol.

Destaining Solution

A general purpose destaining solution, consisting of 5 parts distilled water, 5 parts methyl alcohol, and 1 part glacial acetic acid, is prepared and can be stored for extended periods. This solution is most commonly used in destaining rocket immunoelectrophoresis and crossed immunoelectrophoresis plates.

Appendix D

Normal Ranges of Coagulation Proteins in Plasma

Protein	Concentration
Fibrinogen	200–400 mg/dl
Fibrinopeptide A	0.70–3.1 ng/ml
$B\beta_{15-42}$ peptide	0.43–0.60 pmole/ml
Prothrombin	5–10 mg/dl
Hageman factor	27–45 μg/liter
Factor VIII : RAg	5–10 μg/ml
Protein C	5–10 μg/ml
Fibronectin	290–320 μg/ml
Plasminogen	10–20 mg/dl
α_1-Antitrypsin	250–400 mg/dl
Antithrombin-III	18–30 mg/dl
α_2-Macroglobulin	Males: 150–350 mg/dl
	Females: 175–420 mg/dl
α_2-Antiplasmin	5–6 mg/dl
C_1 Inhibitor	17 mg/dl
Platelet factor 4	2.5–16.0 ng/ml
Beta-thromboglobulin	13–40 ng/ml
TxB_2	0.20 pg/ml

Compiled from reported values in recent literature.

Appendix E

Manufacturers of Immunologic Reagents and Equipment

North American Region

Abbott Laboratories
  Abbott Park
  North Chicago, Illinois 60064 — RIA kit for PF4

Alpha Therapeutic Corporation
  5555 Valley Road
  Los Angeles, California 90032 — Factor VIII concentrate

Amersham Corp.
  2636 Clearbrook Drive
  Arlington Heights, Illinois
    60005 — RIA kit for β-thromboglobulin; ^{125}I fibrinogen analyzers for scanning

Antibodies, Inc.
  P.O. Box 442
  Davis, California 95616 — Antisera to coagulation proteins such as antithrombin-III, factor VIII, etc.

Atlantic Antibodies
  Scarbourough, Maine 04074 — Antisera to coagulation proteins

Baker Instruments Corp.
  2196 Avenue C
  Bethlehem, Pennsylvania 18017 — Laser nephelometers; automated and semiautomated platelet counters; coagulation reagents

Bard Parker
  Rutherford, New Jersey 07070 — Surgical equipment

Biodata
  155 Centennial Plaza
  Horsham, Pennsylvania 19044 — Coagulation profilers; platelet aggregometers; coagulation reagents

BBL
  P.O. Box 243
  Cockeysville, Maryland 21032 — Fibrometer and related products

Bio-Rad Labs
  32nd and Griffin Avenue
  Richmond, California 94804 — Immunoelectrophoresis equipment; reagents for coagulation proteins

Boehringer-Mannheim Biochemicals 7941 Castleway Drive Indianapolis, Indiana 46250	RID plates and antisera
Brinkman Instruments, Inc. Westbury, New York 11590	Electrophoresis equipment
Calbiochem Behring Corp. P.O. Box 12087 San Diego, California 92112	Antisera to coagulation proteins; laser nephelometers; coagulation reagents.
Cal-Med 360 Swift Avenue P.O. Box 2303 San Francisco, California 94080	Chromogenic peptide substrates; plasma and other reagents
FMC Corporation P.O. Box 308 Rockland, Maine 04841	Agarose and gel bind films for electrophoresis
Cappel Laboratories 237 Lacey Street P.O. Box 37 Westchester, PA 19380	Antisera to coagulation proteins; protein standards
CAP 7400 N. Skokie Blvd. Skokie, Illinois 60077	Quality control reagents and survey
Chemical Dynamics Corp. P.O. Box 395 3001 Hadley Road South Plainfield, New Jersey 07080	Standard enzymes; prostaglandin derivatives
Chrono-Log Corp. 2 West Park Road Havertown, Pennsylvania 19803	Luminescence aggregometer; whole blood and other aggregometers
Dako Antibodies 300 Shames Drive Westbury, New York 11590	Antisera to coagulation proteins
Dade Division American Hospital Supply Corp. P.O. Box 672 Miami, Florida 33152	Reagents; Protopath fluorimetric methods for AT-III, plasminogen, heparin, and α_2-antiplasmin control products
Diagnostic Technology, Inc. 290 Community Drive Great Neck, New York 11021	Coagulation calibrators

DuPont Company
  Clinical Systems Division
  Wilmington, Delaware 19898

Fully automated tests for AT-III, plasminogen, and fibrinogen

Electro-Nucleonic
  368 Passaic Avenue
  Fairfield, New Jersey 07006

Centrifugal analyzers for coagulation tests

Ely Lilly Laboratories
  P.O. Box 618
  Indianapolis, Indiana 46202

Protamine sulfate; tranexamic acid; streptokinase–streptodornase; Amicar.

Gelman Sciences, Inc.
  Ann Arbor, Michigan 48106

Immunodiffusion and immunoelectrophoresis templates, gel cutters, buffers, etc.

General Diagnostics
  201 Tabor Road
  Morris Plains, New Jersey
    07950

Bleeding time devices; automated clotting instruments; reagents for coagulation testing; Heparsorb, thrombin time reagent

George King Biomedical, Inc.
  36 Corporate Woods Suite
    3610
  9209 W. 110th Street
  Overland Park, Kansas

Factor deficient substrates (human)

Harvard Apparatus Co, Inc.
  150 Dover Road
  Minis, Massachusetts

Platelet adhesion pump

Helena Laboratories
  P.O. Box 752
  Beaumont, Texas 77704

Electrophoresis; coagulation controls; reference plasmas

Hoechst-Roussel Pharmaceuticals
    Inc.
  Somerville, New Jersey 08876

Streptokinase; Trasylol; enzymes

Hybritech, Inc.
  11085 Torreyana Road
  San Diego, California 92121

Monoclonal antibodies to Factors IX and VIII : C

Hyland Division: Travenol Labs
  2275 Half Day Road
  Deerfield, Illinois

Coagulation reagents; nephelometer

ICL Scientific
  18429 Euclid Street
  Fountain Valley, California
    92708

Rocket electrophoresis equipment; antisera to coagulation proteins

Instrumentation Labs, Inc.
  113 Hartwell Avenue
  Lexington, Massachusetts
    02173

Centrifugal analyzers

International Diagnostics Technology 2551 Walsh Avenue Santa Clara, California 94050	Fluoroimmunoassays (Factor VIII)
Kallestad Laboratories, Inc. 2000 Austin National Bank Tower Austin, Texas 78701	Antisera to coagulation proteins; nephelometric reagents; other control reagents
Mallinckrodt, Inc. 675 McDonnell Boulevard P.O. Box 5840 St. Louis, Missouri 63134	RIA kits for FPA
Mico USA, Inc. 180 Riverside Drive New York, New York 10024	RIA kits for Bβ15-42
Nordic Immunological Laboratories P.O. Box 3715 El Toro, California 92630	Immunological reagents for coagulation analysis
Ortho Diagnostics Raitan, New Jersey 08869	Coagulation reagents
Pacific Hemostasis 4100 Easton Drive Bakersfield, California 93309	Coagulation reagents; factor deficient plasmas; venoms
Pel-Freez Biologicals P.O. Box 68 Rogers, Arkansas 72756	Blood products; agarose preparations
Payton & Associates 244 Delaware Avenue Buffalo, New York 14202	Platelet aggregometers
Peninsula Laboratories P.O. Box 1111 San Carlos, California 94070	Antibodies to coagulation proteins
Pharmacia Fine Chemicals 800 Centennial Avenue Piscataway, New Jersey 08854	Coagulation reagents and kits; heparin–Sepharose gel for AT-III purification; lysine–Sepharose for plasminogen purification
Scientific Products 1430 Waukegan Road McGaw Park, Illinois 60085	Distribution for coagulation reagents from many different manufacturers
Sherwood Medical Industries 1831 Olive Street St. Louis, Missouri 63178	Automated coagulation instruments; coagulation reagents and supplies

Sienco, Inc.
  P.O. Box 108
  Star Route
  Morrison, Colorado 80465

Platelet aggregometer; coagulation reagents; Sonoclot instrument

Sigma Chemicals Co.
  P.O. Box 14508
  St. Louis, Missouri 63178

Coagulation reagents and kits for heparin and AT-III; luciferase–luciferin for platelet aggregation

Syva Corp.
  3181 Porter Drive
  Palo Alto, California 94304

EMIT, ELISA technology

Technidyne Corp.
  138 Forrest Street
  Metuchen, New Jersey 08840

Hemochron timers; activated coagulation tests; heparin monitoring

Vega Biochemicals
  P.O. Box 11648
  Tucson, Arizona 85734

FSP; protein standards

Vol-u-Sol
  1034 S. Commerce Street
  Las Vegas, Nevada 89106

Thrombelastograph; coagulation reagents; nephelometer

Wellcome Reagents
  Division of Burroughs
    Wellcome Co.
  3030 Cornwallis Road
  Research Triangle, North Carolina 27709

Fibrin degradation product kits

European Region*

Alpha Laboratories
  40 Parhan Drive
  Eastleigh, Hampshire
  Great Britain S05 4NU

Disposables; antisera

Armour Pharmaceuticals
  Eastbourne, East Sussex
  Great Britain BN22 OAG

Blood products (BSA)

Associated Hospital Supply
  P.O. Box 4
  Pershore, Worcestershire
  Great Britain

Antisera [agent for Kallestad]

---

* Compiled from Thompson RA (ed): Techniques in Clinical Immunology. Oxford, Blackwell Scientific, 1981.

Beckman Instruments
  Beckman RIIC Ltd.
  Cressex Industrial Estate
  Turnpike Road
  High Wycombe, Buckingham-
    shire
  Great Britain
	Nephelometers; immunologic re-
	agents

Bio Merieux
  Marcy L'etoile, 69260
  Charbonnières-les-Bains
  France
  [Laboratory Impex–UK]
	Screening kits; antisera

Bio-Rad Laboratories
  Caxton Way
  Watford, Hertfordshire
  Great Britain
	Electrophoresis immunoreagents

Biotext Folex Ltd.
  171 Alcester Road
  Moseley, Birmingham
  Great Britain B13 8JR
	Antisera

Boehringer Corp.
  Boehringer Mannheim GmbH
  6800 Mannheim 31
  West Germany
	Antisera; biochemicals

Boehringer Corp., (Bilton House)
  Uxbridge Road
  London W5
  Great Britain
	Antisera; biochemicals

CIS (UK) Ltd.
  Rex House
  354 Ballards Lane
  North Finchley, London
  Great Britain N12 OEG
	Radioimmunoassay kits

Diamed Diagnostics
  Mast House
  Derby Road
  Bootle, Merseyside
  Great Britain L20 1EA
	Hemagglutination [agent for Fuji-
	zoki]

Dynatech
  Daux Road
  Billingshurst, Sussex
  Great Britain
	Microtitration equipment; antisera

Flow Laboratories P.O. Box 17 Heatherhouse Road Irvine, Ayrshire Great Britain KA12 8N8	Antisera: microtitration appliances
Gelman Hawksley 10 Harrowden Road Brackmills, Northampton Great Britain NN4 OEB	Electrophoresis apparatus
Gerrard T. & Co. Worthing Road East Preston, Littlehampton Great Britain	Animal suppliers
Gibco Europe Ltd. 3 Washington Road Sandyford Estate, Paisley Great Britain PA3 4EP	Media; agent Meloy; Sylvania antisera
Helena Laboratories MI Scientific Ltd. Exchange Building Quayside, Newcastle-upon-Tyne Great Britain NE1 3BJ	Electrophoresis; immunologic reagents
Hoechst Pharmaceutical Hoechst House Salisbury Road Hounslow, Middlesex Great Britain TW4 6JH	Antisera: electrophoresis; RID plates
ICL Scientific Ltd. Boehringer Corp. Bell Lane Lewes, East Sussex Great Britain BN7 1LG	Antisera
Immuno Ltd. Immuno Diagnostika Immuno AG A-1220 Vienna, Austria Immuno Diagnostica Rye Lane, Dunton Green Seven Oaks, Kent Great Britain TN14 5HB	Antisera; blood products
IMCO Corp. Lt. Hrdiksvaklis Gatan 4B 113-36 Stockholm, Sweden	RIA kits

Institutes Pasteur 36 Rue de Docteur Roux Paris 15, France [Uniscience—UK]	Antisera
International Enzymes Ltd. Vale Road Windsor, Berkshire Great Britain S14 5NLJ	Bioreagents
Kallestad Kabi AB Stockholm, Sweden [Associated Hospital Supply— UK]	Antisera; immunoreagents; kits; Substrates
Koch Light Laboratories Ltd. Poyle Estate Willow Broad Colnbrook Slough, Berkshire Great Britain SL3 OBZ	Biochemicals; scintillants
Kontron-Interechnique Kontrol House P.O. Box 88 Campfields Road St. Albans, Hertfordshire Great Britain AL1 5JG	Scintillation vials
Laboratoire Roger Bellon 159 Avenue de Poule Neuilly, France [Uniscience—UK]	Blood plasma substitutes
Laboratory Impex Ltd. Lion Road Twickenham, Middlesex Great Britain TW1 4ZF	Antisera
MI Scientific Exchange Building Quayside, Newcastle-upon- Tyne Great Britain NE1 3BJ	Electrophoresis [Helena products]
Marine Colloids [Uniscience—UK]	Agar; agarose
Meloy [Gibco—Europe—UK]	Antisera

Miles Laboratories   Agar; antisera
  P.O. Box 37
  Stoke Poges, Slough
  Great Britain SL2 4LY
  [Miles Seravac, Miles Yeda]

Millipore   Cellulose membranes; electropho-
  Abbey Road, London   resis
  Great Britain NW10 7SP

New England Nuclear (NEN)   Radionuclides
  Dorval, Quebec
  Canada

NEN Chemicals GmbH   RIA kits
  37 Windermere Liden
  Swindon, Wiltshire
  Great Britain

Nordic Immunological Laborato-   Antisera
    ries
  P.O. Box 22
  Tilburg, The Netherlands

Nordic Immunological Laborato-   Antisera and standard plasma
    ries
  60 King Street
  Maidenhead, Berkshire
  Great Britain SL6 1EQ

North East Laboratories   Antisera; microbial agents
  Biomedical Laboratories Ltd.
  P.O. Box 45
  Uxbridge, Middlesex
  Great Britain UB9 5QD
  [Alpha]

Nyegaard & Co.   Antisera; reagents
  A.S. Oslo
  Nycorn 2
  Oslo 4, Norway
  [Uniscience—UK]

Organon Teknika   Antisera
  Industrielaan 84
  Oss, Holland

  Teknika House   Antisera; standard reagents
  Cromwell Road
  St. Neots
  Huntingdon, Cambridgeshire
  Great Britain PE19 2EU

Ortho Diagnostics
  Denmark Street
  High Wycombe, Buckingham-
    shire
  Great Britain HP11 2ER

Screening kits; antisera

Oxoid Ltd.
  Wade Road
  Basingstoke, Hampshire
  Great Britain RG24 OPW

Agar; cellulose membranes

PCD Ltd.
  42 Queens Road
  Farnborough, Hampshire
  Great Britain

2-dimensional electrophoresis
  (timed power pack)

RIA (UK) Ltd.
  3 Manor Place
  Athenaeum Street
  Sunderland
  Great Britain SR1 1QX

RIA kits

Radiochemical Centre
  White Lion Road
  Amersham, Buckinghamshire
  Great Britain 4P7 9LL

Radionuclides

Roche Products Ltd.
  P.O. Box 8
  Welwyn Garden City,
    Hertfordshire
  Great Britain AL7 3AY

RIA kits

Sera Laboratories Ltd.
  Crawley Down, Sussex
  Great Britain RH10 4LL

Animals; antisera

Serva Fine Biochemicals
  P.O. Box 105260
  Karl Benz Strasse 7
  D6900 Heidelberg 1
  West Germany
  [Uniscience—UK]

Electrophoretic supplies

Seward Laboratory
  UAC House
  Blackfriars Road
  London
  Great Britain SE1 9UG

Antisera

Shandon Southern Instruments Ltd.   Electrophoresis; chromatography
  Astmoor Industrial Estate
  Runcorn, Cheshire
  Great Britain WA7 1PR

Sigma (London) Chemical Co.   Biochemicals; kits
  Fancy Road
  Poole, Dorset
  Great Britain BH17 7NH

TAAB Laboratories   Chemicals (e.g., glutaraldehyde)
  52 Kidmore End Road
  Emmer Green
  Reading, Berkshire
  Great Britain

Tissue Culture Services   Animal products; antisera
  10 Henry Road   (TAGO)
  Slough
  Great Britain SL1 2QL

Travenol   Antisera; blood products; laser
  Caxton Way
  Thetford, Norfolk
  Great Britain IP24 38E

Uniscience Ltd.   Agents for multiple products
  Uniscience House
  8 Jesus Lane
  Cambridge
  Great Britain C85 8BA

Wellcome Reagents Ltd.   Immunodiagnostics
  303 Hither Green Lane
  London
  Great Britain SE13 6TL

Appendix F

Glossary of Terms

Adjuvant. Substance injected with an antigen in order to enhance the immune response.

Affinity. Binding force between antigen and antibody in an antigen–antibody reaction.

Afibrinogenemia. Absence of fibrinogen in plasma.

Agglutination. A type of antigen–antibody reaction in which a particular antigen forms a lattice with its monospecific antibody.

Agglutinin. Antibody specific for surface antigens.

Allogenic. Belonging to different genetic or antigenic type.

Antibody. Serum protein (gamma globulin) induced by an antigen and which can react specifically with the antigen.

Antibody binding site. Portion of antibody molecule which combines with the corresponding antigenic determinants.

Antigen. Macromolecule which will induce antibody response when injected into an animal.

Antigenic determinant. Area of an antigen that determines specificity of the immune reaction.

Antigenicity. Ability of a substance to react with an antibody.

Antiserum. Serum containing an antibody.

Avidity. Degree of binding between an antibody and an antigen.

Blood group antigens. Genetically determined antigens present on red blood cell surface.

Constant region. Carboxyl terminal portion of an antibody which is identical in immunoglobulin molecules of a given class.

C-reactive protein. Protein secreted by the liver during an inflammatory disease.

Cross-reaction. Reaction of an antibody with a closely related antigen.

Cryoprecipitate. Cold-induced precipitate of normal human plasma which is rich in antihemophilic factor.

Disseminated intravascular coagulation. A clinicopathologic syndrome in which simultaneous activation of coagulation and the fibrinolytic pathways takes place.

Diversity.   Presence of different antigen-specific combining sites.

Dysfibrinogenemia.   A familial disorder of functionally abnormal fibrinogen.

Fibrinolysis.   A process of enzymatic degradation of fibrin.

Fluorescence.   Emission of light of one color when a substance is irradiated with light of a different color.

Freund's complete adjuvant.   An emulsion which contains killed mycobacteria; it is used with an antigen to enhance the immune response.

Gamma globulins.   Serum proteins migrating at a gamma position during electrophoresis. Includes immunoglobulins.

Hapten.   A substance which is not immunogenic, but can react with a specific antibody.

Hemagglutination inhibition.   Inhibition of red cell agglutination by a homologous antigen. Red cells are coated with a specific antibody.

Hemophilia.   An inherited disorder with a tendency to bleed due to coagulation factor deficiency. In hemophilia A, Factor VIII is deficient; hemophilia B is due to Factor IX deficiency.

Hemostasis.   Hemostasis refers to the arrest of bleeding upon an injury.

Heparin.   A mucopolysaccharide which has an anticoagulant effect on hemostasis.

Heparin cofactor.   An α_2-globulin (antithrombin-III) required for the anticoagulant action of heparin.

Hybridoma.   Hybrid cells secreting specific antibodies. These cells are derived from the fusion of plasma cells with myeloma cells.

Immune-complex disease.   Disease caused by antigen–antibody complexes; e.g., glomerulonephritis.

Immunity.   Capacity to become resistant to an infection.

Immunogen.   Any substance capable of producing an immune response when injected into an animal.

Myeloma.   Excessive production of one or more immunoglobulins.

Neoantigen.   New antigen produced as a result of antigenic modification.

Specificity.   A selective reaction between an antigen and an antibody.

Thrombasthenia.   An inherited disorder of platelet function.

Thrombocytopenia.   A condition in which platelet count is decreased in circulating blood.

Thrombocytosis.   An increase in the number of platelets in circulating blood.

Thrombosis.   Formation of a blood clot within a vessel.

Thrombus.   A blood clot.

von Willebrand's disease.   An inherited bleeding disorder in which the concentration of both Factor VIII : RAg and Factor VIII : CAg is low.

References

1. Abramson N, Schur PH: The IgG subclasses of red cell antibodies and relationship to monocyte binding. Blood 40:500–508, 1972.

2. Aggler PM, White SG, Glendenning MB, Page EW, Leake TB, Bates G: Plasma thromboplastin component (PTC) deficiency: a new disease resembling hemophilia. Proc Soc Exp Biol Med 79:692–694, 1982.

3. Alexander RL: Comparison of radial immunodiffusion and laser nephelometry for quantitating some serum proteins. Clin Chem 26:314–317, 1980.

4. Allen RM, Redshaw MR: The use of homologous and heterologous ^{125}I-radioligands in the radioimmunoassay of progesterone. Steroids 32:467–486, 1978.

5. Arkin CF, Hartman AS: The hypercoagulability states. In Batsaki J, Savory J (eds): Critical Reviews in Clinical Laboratory Science. Baton Rouge, Fl., CRC Press, 1979.

6. Baugh RF, Hongie C: The chemistry of blood coagulation. Clin Haematol 8(1):3–30, 1979.

7. Bick R: Disseminated intravascular coagulation (DIC) and related syndromes. In Fareed J, Messmore HL, Fenton JW, Brinhous KM (eds): Perspectives in Hemostasis. New York, Pergamon, 1981, pp 122–138.

8. Biggs R, Douglas AS, Mcfarlane RG, Dacie JV, Pitney WR, Merskey C, O'Brien JR: Christmas disease, a condition previously mistaken for hemophilia. Br Med J 2:1378–1382, 1952.

9. Bloch KJ, Maki DG: Hyperviscosity syndromes associated with immunoglobulin abnormalities. Semin Hematol 10:113–124, 1973.

10. Bolwerk CJM, Henny CP, Buller HR, Ten Cate JW: Hereditary AT-III deficiency, and heparin. Thromb Res 28:689–690, 1982.

11. Brown JE, Carton CL, Hougie C: The effect of naturally occurring antibodies to factor VIII on an immunoradiometric assay for factor VIII coagulant antigen. Observation in a cross-reacting material-positive (CRM$^+$) hemophiliac with a factor VIII inhibitor. J Lab Clin Med 97(1):65–71, 1981.

12. Buckley RH, MacQueen JM, Ward FE: HLA antigen in primary immunodeficiency disease. Clin Immunol Immunopathol 7:305–310, 1977.

13. Buffone GJ, Savory J, Cross RE: Use of a laser-equipped centrifugal analyzer for kinetic measurement of serum IgG. Clin Chem 20(10):1320–1323, 1974.

14. Buffone GJ, Savory J, Cross RE, Hammond JE: Evaluation of kinetic light scattering as approach to the measurement of specific proteins with the centrifugal analyzer. I. Methodology. Clin Chem 21(12):1731–1734, 1975.

15. Carpenter CB (ed): Clinical Histocompatibility Testing. A Transplantation Proceeding Reprint, Vol. 2. New York, Grune & Stratton, 1977.

16. Cejka J, Kithier K: A simple method for the classification and typing of monoclonal immunoglobulins. Immunochemistry 13:629–631, 1976.

17. Chang JJ, Crowl CP, Schneider RS: Homogenous enzyme immunoassay for digoxin. Clin Chem 21:967, 1975.

18. Clemmensen I: Different molecular forms of α_2-antiplasmin. In Collen D, Wiman B, Verstraete M (eds): The Physiological Inhibitors of Coagulation and Fibrinolysis. Amsterdam, Elsevier/North-Holland, 1979, pp 131–136.

19. Counts RB: Solid phase immunoradiometric assay of factor VIII protein. Br J Haematol 31:429–436, 1975.

20. Curry RE, Hertzman H, Reige DH, Sweet RV, Simonsen MG: A systems approach to fluorescent immunoassay: general principles and representative applications. Clin Chem 25:1591–1595, 1979.

21. Dash RJ, England BG, Midgely AR, Niswender GD: A specific, nonchromatographic radioimmunoassay for human plasma cortisol. Steroids 26:647–661, 1975.

22. Davie EW, Fujikawa K, Kuvachi K, Kisiel W: The role of serine proteases in the blood coagulation cascade. Adv Enzymol 49:277–311, 1979.

23. Davie EW, Kirby EP: Molecular mechanisms in blood coagulation. Curr Top Cell Reg 7:51–86, 1973.

24. Denson K, Lurie WE, De Cataldo F, Mannucci PM: The factor-X deficit: recognition of abnormal forms of factor-X. Br J Haematol 18:317–327, 1970.

25. Denton CD, Kameron WM, Smith RS, Creveling RL: Use of laser nephelometry in the measurement of serine proteins. Clin Chem 22(9):1465–1471, 1976.

26. Dito WR: Rapid immunonephelometric quantitation of eleven serum proteins by centrifugal fast analyzer. Am J Clin Pathol 17:301–308, 1979.

27. Dyas J, Read GF, Fahmy D: A simple robust assay for testosterone in male plasma using an ^{125}I-radioligand and a solid-phase separation technique. Ann Clin Biochem 61:325–330, 1979.

28. Ebring R, Schmidt W, Fuchs G, Havemann K: Demonstration of granulocytic proteases in plasma of patients with acute leukemia and septicemia with coagulation defects. Blood 49(2):219–321, 1977.

29. Fahey JL, McKelvey EM: Quantitative determination of serum immunoglobulins in antibody-agar plates. J Immunol 94:84–90, 1965.

30. Fahmy DR, Read GF, Hillier SG: Some observations on the determination of cortisol in human plasma by radioimmunoassay using antisera against cortisol-3-BSA. Steroids 26:267–280, 1975.

31. Fantl P, Sawers RJ, Marr AG: Investigation of haemorrhagic disease due to beta-prothromboplastin deficiency complicated by a specific inhibitor of thromboplastin formation. Aust Ann Med 5:163–176, 1956.

32. Finley PF, Williams RJ, Byers JM: Immunochemical determination of human immunoglobulins with a centrifugal analyzer. Clin Chem 22(7):1037–1041, 1976.

33. Fubara ES, Freter R: Protection against enteric bacterial infection by secretory IgG antibodies. J Immunol 3:395–403, 1973.

34. Gaffney PJ, Joe FE, Mahmoud M, Fossati CA, Spitz M: A novel radioimmunometric approach to the study of components of human hemostasis. I. Assay of plasma fibrinopeptide A levels. Thromb Res 19(6):815–822, 1980.

35. Gamba G, Fornasari P, Montani N, Biancardi M, Griganani B, Ansari E: Plasma levels of protease inhibitors in acute myeloid leukemia at the onset of the disease and during antiblastic therapy. Thromb Res 17:41–53, 1980.

36. Girma JP, Ardaillou N, Meyer D, Lavergne JM, Larrieu MJ: Fluid phase immunoradiometric assay for the detection of qualitative abnormalities of factor VIII from Willebrand factor in variants of von Willebrand's disease. J Lab Clin Med 93:926–939, 1979.

37. Girma JP, Lavergne JM, Meyer D, Larrieu MJ: Immunoradiometric assay of factor VIII: Coagulation antigen using four human antibodies, study of 27 cases of hemophilia A. Br J Haematol 47(2):269–282, 1981.

38. Girolami A, Bareggi G, Brunetti A, Sticchi A: Prothrombin Padua: a "new" congenital dysprothrombinemia. J Lab Clin Med 84:654, 666, 1974.

39. Goldman M: Fluorescent antibody methods. New York, Academic Press, 1968.

40. Grabar P, Williams CT: Methode permettant l'etude conjugee des proteines. Application au serum sanguim. Biochem Biophys Acta 10:193–194, 1953.

41. Gralnick HR, Coller BS, Frantantoni JC, Martinez J: Fibrinogen Bethesda III: a hypodysfibrinogenemia. Blood 53(1):28–46, 1979.

42. Gueffroy DE: Buffers: a guide for the 1978 preparation and use of buffers in biological systems. Calbiochem-Behring Corporation, San Diego, 1981.

43. Harpel PC, Rosenberg RD: α_2-Macroglobulin and antithrombin-heparin cofactor: modulators of hemostatic and inflammatory reactions. In Spaet TH (ed): Progress in Hemostasis and Thrombosis. New York, Grune & Stratton, 1976, pp 145–190.

44. Heidelberger M, Kendal FE: A quantitative study of the precipitin reaction between type III pneumococcus polysaccharide and purified homologous antibody. J Exp Med 50:809–823, 1929.

45. Highsmith RF, Rosenberg RD: The inhibitions of human plasmin by human antithrombin-heparin cofactor. J Biol Chem 249:4335–5338, 1974.

46. Howard MA, Firkin BA: Ristocetin—a new tool in the investigation of platelet aggregation. Thromb Diath Haemorrh 26:362–369, 1971.

47. Hoyer IW: Immunological studies of antihemophilic factor (AHF, factor VIII). IV. Radioimmunoassay of AHF antigen. J Lab Clin Med 80:822–833, 1972.

48. Hutlin MV, London FS, Shapiro SS, Young WJ: Heterogeneity of factor VIII antibodies: further immunochemical and biologic studies. Blood 49:807–817, 1977.

49. Joyce BG, Fahmy D, Hillier SG: Specific determination of testosterone in female plasma by radioimmunoassay: a rapid and reliable procedure for the routine clinical laboratory. Clin Chim Acta 62:231–238, 1975.

50. Kass L, Ratnoff OD, Leon MA: Studies on the purification of antihemophilic factor (factor VIII). I. Precipitation of antihemophilic factor by concanavalin-A. J Clin Invest 48:351–358, 1969.

51. Killingsworth LM, Savory J: Automated immunochemical procedures for measurement of immunoglobulins IgG, IgA, and IgM in human serum. Clin Chem 17:936, 940, 1971.

52. Killingsworth LM, Savory J: Manual nephelometric methods for immunochemical determination of immunoglobulins IgG, IgA, and IgM in human serum. Clin Chem 18(4):335–339, 1972.

53. Killingsworth LM, Savory J: Measurement of immunoglobulins in cerebrospinal fluid employing nephelometric immunoprecipitin techniques. Clin Chim Acta 43:279–281, 1973.

54. Killingsworth LM, Savory J: Nephelometric studies of the precipitin reaction: a model system for specific protein measurements. Clin Chem 19(4):403–407, 1973.

55. Kisiel W, Hanahan DJ: Purification and characterization of human factor II. Biochim Biophys Acta 304:103–113, 1973.

56. Kluft C, Los N: Demonstration of two forms of α_2-antiplasmin by modified crossed immunoelectrophoresis. Thromb Res 21:65–71, 1981.

57. Koehler G, Milstein C: Derivation of specific antibody–producing tissue culture and tumor lines by cell fusion. Eur J Immunol 6:511–519, 1976.

58. Koutts J, Meyer D, Rickard K, Scott L, Firkin BG: Heterogeneity in biological activity of human factor VIII antibodies. Br J Haematol 29:99–107, 1975.

59. Kuusi N: A technical improvement for crossed immunoelectrophoresis. J Immunol Methods 31:361–364, 1979.

60. Lanchantin GF, Hart DW, Friedman AJ, Saavedra NV, Mehl JW: Amino acid composition of human plasma prothrombin. J Biol Chem 243:5479–5485, 1968.

61. Lane JL: Some immunological investigations on antithrombin III "Budapest." Br J Hematol 40:459–470, 1978.

62. Langone JL, Vunakis HV (eds): Immunochemical Techniques. Methods Enzymol 84:51–60, 1982.

63. Laurell CB: Antigen–antibody crossed electrophoresis. Anal Biochem 10:358–361, 1965.

64. Legaz ME, Schmer G, Counts RB, Davie EW: Isolation and characterization of human factor VIII (antihemophilic factor). J Biol Chem 248:3946–3955, 1973.

65. Leute RK, Ullman EF, Goldstein A, Herzen LA: Spin immunoassay technique for determination of morphine. Nature (New Biol) 236:93–94, 1972.

66. Levann M, Rimon S, Shari M, Ramot S, Goldberg E: Active and inactive factor VII in Dubin Johnson syndrome with factor VII deficiency and in Coumadin administration. Br J Haematol 23:669–677, 1972.

67. Levy AL: Fluorescent immunoassays for therapeutic drugs. In Sunderman FW (ed): Applied Seminar on Laboratory Diagnosis of Disorders of the Fetus, Newborn, and Infant. Philadelphia, Institute for Clinical Science, 1981, pp 97–100.

68. Ljung R, Holmberg L: Factor VIII : CAg in hemophilia A. A comparison between IRMA : S using hemophilic and spontaneous antibodies. Thromb Res 24:45–50, 1981.

69. Macfarlane RG: An enzyme cascade in the blood clotting mechanism, and its function as a biochemical amplifier. Nature 202:498–499, 1964.

70. Magnusson S: Primary structure studies on thrombin and prothrombin. Thromb Diath Haemorrh 54:31–35, 1973.

71. Mammen EF (ed): Congenital coagulation disorders. Semin Thromb Hemostasis 9(1):1–72, 1983.

72. Mammen EF, Prasas AS, Barnhart MI, Au CC: Congenital dysfibrinogenemias: molecular abnormalities of fibrinogen. Blut 33:229–234, 1976.

73. Mancini G, Carbonara AO, Heremans JF: Immunochemical quantitation of antigens by single radial immunodiffusion. Immunochemistry 2:235–254, 1965.

74. Marder VJ: The functional defects of hereditary dysfibrinogens. Thromb Hemostasis (Stuttgart) 36:1–8, 1976.

75. Messmore HL: Natural inhibitors of the coagulation system. Semin Thromb Hemostasis 8(4):267–275, 1982.

76. Morse EE: The fibrinogenopathies. Ann Clin Lab Sci 8(3):234–238, 1978.

77. Movat HZ, Ozge-Anwar AH: The contact phase of blood coagulation. Clotting factors XI and XII: their isolation and interactions. J Lab Clin Med 84:861–878, 1974.

78. Muller HP, van Tilburg NH, Bertina RM, Terwiel JP, Veltkamp JJ: Immunoradiometric assay of procoagulant factor VIII antigen (VIII CAg). Clin Chem Acta 107(1–2):11–19, 1980.

79. Nilsson IM: Report of the working party on factor VIII related antigens. Thromb Hemostasis 39:511–520, 1978.

80. Oserud B, Miller-Andersen M, Abilgaard V, Prydz A. The effect of antithrombin-III on the activity of the coagulation factors VII, IX, and X. Thromb Hemostasis 35:295–304, 1976.

81. Ouchterlony O: Antigen–antibody reactions in gel. Ark Kem Min Geol 26B:1, 1948.

82. Owen CA: Hemostasis—past, present, and future. Mayo Clin Proc 55:505–508, 1980.

83. Pal SB: Enzyme-labelled Immunoassay of Hormones and Drugs. Berlin, de Gruyter, 1978.

84. Parvez Z, Fareed J, Messmore HL, Bermes EW: Laser and rate nephelometric methods for the quantitation of coagulation proteins. In Manual of Procedures for the Seminar on Biochemical Hematology, Sunderman FW (ed): Institute for Clinical Sciences Inc, Philadelphia, PA, 1979.

85. Parvez Z, Messmore HL, Seghatchian MJ: Detection of complexes between serine proteases from leukemic cells and serine protease inhibitors. Blood 60(5):134a, 1982.

86. Peake FR, Bloom AL, Giddings JC, Ludlam CA: An immunoradiometric assay for procoagulant factor VIII antigen: results in hemophilia, von Willebrand's disease, and fetal plasma and serum. Br J Haematol 42(2):269–281, 1979.

87. Petersen TE, Dudek-Nojcienchowska G, Sottrup-Jensen L, Magnusson S: The primary structure of antithrombin-III. Thromb Hemostasis 38:201, 1977.

88. Peterson JW, Hetjtmancik KE, Markel DF, Craig JP, Kurosky A: Antigenic specificity of neutralizing antibody to cholera toxin. Infection Immunity 24:774–779, 1979.

89. Prydz HH, Gladhaug A: Factor X immunological studies. Thromb Diath Haemorrh 25:157–165, 1971.

90. Rabiet MJ, Elion J, Benarous R, Labie D, Josso F: Activation of prothrombin Barcelona: evidence for active high molecular weight intermediates. Biochim Biophys Acta 584:66–75, 1979.

91. Ratnoff OD: Antihemophilic factor (factor VIII). Ann Intern Med 88:403–409, 1978.

92. Ressler RN: Electrophoresis of serum protein antigens in an antibody containing buffer. Clin Chim Acta 5:359–365, 1960.

93. Rick ME, Hoyer LW: Immunologic studies of anti-hemophilic factor (AHF) (factor VIII). V. Immunologic properties of AHF subunits produced by salt dissociation. Blood 42:737–747, 1973.

94. Ritchie RF, Alper CA, Graves J, Pearson N, Larson C: Automated quantitation of proteins in serum and other biologic fluids. Am J Clin Pathol 59:151–159, 1973.

95. Ritzman SE: Radial immunodiffusion revisited. Lab Med 9(7):23–33, 1978.

96. Rosenberg RD, Damus PS: The purification and mechanism of action of human antithrombin: heparin cofactor. J Biol Chem 248:6490–6505, 1973.

97. Sas G, Peto F, Banghegyi D, Blasko G. Domjan G: Heterogeneity of the "classical" antithrombin-III deficiency. Thromb Hemostasis 43:133–136, 1980.

98. Schenkel-Brunner H, Catron JP, Doinel C: Localization of blood-group A and 1 antigenic sites on inside-out and right side–out human erythrocyte membrane vesicle. Immunology 36:33–36, 1979.

99. Schultze HE, Schwick G: Quantitative immunologische Bestimmung von Plasmaprotenen. Clin Chem Acta 4:15–25, 1959.

100. Schwab JH: Suppression of the immune response by microorganisms. Bacteriol Rev 39:121–143, 1975.

101. Scully MF, Deltan H, Chan P, Kakar VV: Hereditary antithrombin-III deficiency in an English family. Br J Haematol 47:235–240, 1981.

102. Seegers WH: A personal perspective on hemostasis and thrombosis (1937–1981). Semin Thromb Hemostasis 7(3–4):180–198, 1981.

103. Shanberge JKN, Gore I: Studies on the immunologic and physiologic activities of antihemophilia factor (AHF). J Lab Clin Med 50:954, 1957.

104. Shapiro S, Martinez J, Holburn RR: Congenital dysprothrombinemia: an inherited structural disorder of human prothrombin. J Clin Invest 48:2251–2259, 1969.

105. Shewhart WA: Economic Control of Quality of Manufactured Products. New York, Van Nostrand, 1931.

106. Sorensen PJ, Dyerberg J, Stoffersen E, Jensen MK: Familial functional antithrombin-III deficiency. Scand J Haematol 24:105–109, 1980.

107. Soria J, Soria C, Ryckewaert JJ: A solid phase immunoenzymological assay for the measurement of human fibrinopeptide A. Thromb Res 20:425–535, 1980.

108. Sternberg JC: A rate nephelometer for measuring specific proteins by immunoprecipitin reactions. Clin Chem 23(8):1456–1464, 1977.

109. Tanaka M, Kato K: Determination of antithrombin-III by sandwich enzyme immunoassay technique. Thromb Res 22:67–74, 1981.

110. Tanaka M, Minesaki M, Kato K: Measurement of fibrinogen degradation products by a sandwich enzyme immunoassay technique. Clin Chim Acta 103:287–295, 1980.

111. Thomas J, Thomas F, Mendez-Picon G, Lee H: Immunological monitoring of long-surviving renal transplant recipient. Surgery 81:125–131, 1977.

112. Thompson JM (ed): Blood Coagulation and Haemostasis. A Practical Guide, 2nd ed. Edinburgh, Churchill Livingstone, 1980.

113. Tiffany TO, Parella JM, Johnson WF, Burtis CA: Specific protein analysis by light scatter measurement with a miniature centrifugal fast analyzer. Clin Chem 20(8):1055–1061, 1974.

114. Triplett DA: New methods in coagulation. Crit Rev Clin Lab Sci 15(1):25–84, 1981.

115. Triplett DA, Harms CS: Procedures for the Coagulation Laboratory. Chicago, American Society of Clinical Pathologists, 1981.

116. Voller A, Bartlett A, Bidwell D (eds): Immunoassays for the 80's. Lancaster, MTP Press, 1981.

117. Watkins WM: Blood group substances. Science 152:172–181, 1966.

118. Westgard JO, de Vos DJ, Hunt MR, Quam EF, Carey RN, Garber CC: Method Evaluation. Bellaine, American Society of Medical Technology, 1978.

119. Yalow RS, Berson SA: Immunoassay of endogenous plasma insulin in man. J Clin Invest 39:1157–1175, 1960.

120. Youder JM, Schick LA, Moore RP: A convenient, rapid fluoroimmunoassay for factor VIII related antigen. Thromb Res 24:51–59, 1981.

121. Youden WJ: Statistical Methods for Chemists. New York, Wiley, 1951.

122. Zdebska E, Koscielak J: Studies in the structure and I blood-group activity of poly (glycosyl) ceramides. Eur J Biochem 91:517–525, 1978.

123. Zimmerman TS, Abildgarrd C, Meyer D: The factor VIII abnormality in severe von Willebrand's disease. N Engl J Med 301:1307–1310, 1979.

124. Zimmerman TS, Ratnoff OD, Powell AE: Immunologic differentiation of classic hemophilia (factor VIII deficiency) and von Willebrand's disease with observations in combined deficiencies of antihemophilic factor and proacelerin (factor V) and in acquired circulating anticoagulant against antihemophilia factor. J Clin Invest 50:244–254, 1971.

Index

Everything should be made as simple as possible, but not one bit simpler.

—Albert Einstein